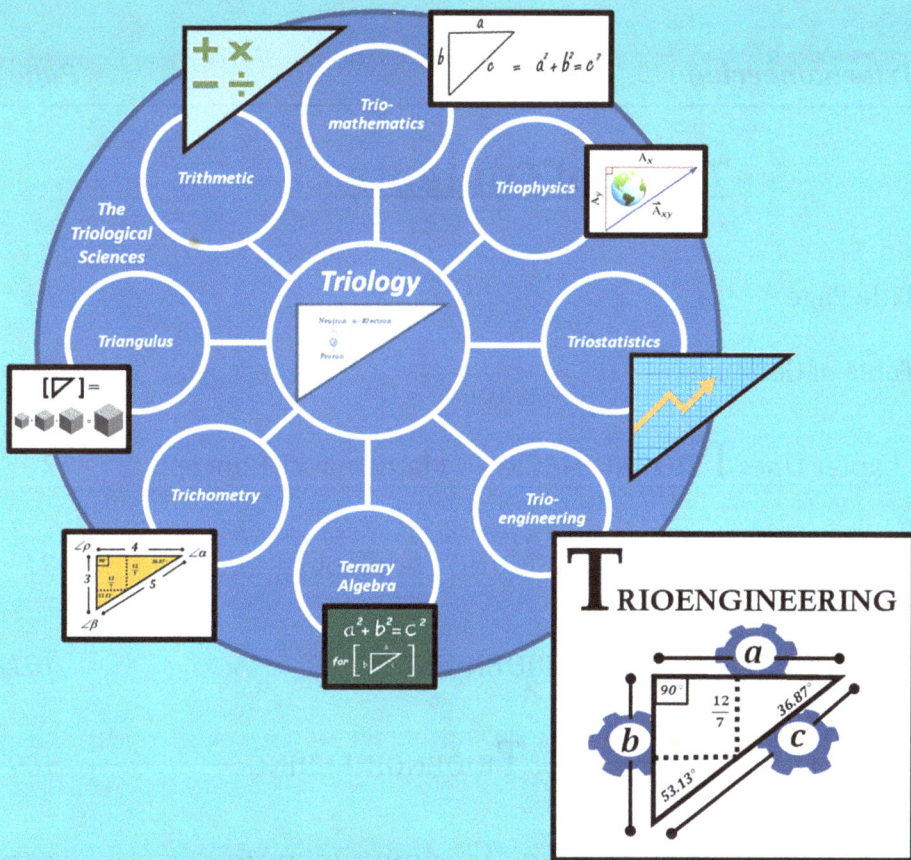

TRIOENGINEERING ™ ©

The In-Depth Trichotomous Science of
the Dynamic 3-4-5-6 Golden Upright Right
Triangle for Innovative Problem-Solving

By

James E. Osler II, Ed.D.

Table of Contents

Library of Congress Cataloging–in–Publication Data

Osler II, James E.

TRIOENGINEERING ™ © *The Problem-Solving Triological Science:*
The In-Depth Trichotomous Science of the Dynamic 3-4-5-6 Golden
Upright Right Triangle Volumetrics for Innovative Problem-Solving. 1st
Edition.

ISBN: 978–1–938978–16–8

1. English Language–Mathematics. 2. Innovation, 3. Measurement, and
4. Triological Sciences

Published in the United States.

ISBN: 978–0–9817189–16–8

Dedication and Special Acknowledgements

Dedication

I send a great and appreciative "Thank You" to Almighty GOD for his continual blessings and for His son my Lord and Savior Jesus Christ, through whom all things are possible. This book is designed to bless all who read it and provide them with greater insight and understanding. Oh Lord, be thy glorified in all that I do.

Special Acknowledgements

I send a very special "Thank You" to my family. Your support has greatly aided me in seeing this project completed to the very end. You are a blessing. Thank you and I love you all. An extra special "Thank You" to Dr. Philliph Masila Mutisya who has been an ongoing support throughout my career in higher education and a critical and essential component to my academic growth as a professional and as a professor.

About the Author

A native of North Carolina, James Edward Osler II was born and raised in the City of Medicine, Durham, NC. An artist, Osler is always using art to teach in some form or capacity. He completed a Baccalaureate in Art at North Carolina Central University (NCCU) with a concentration in Studio emphasizing painting, illustration, and design.

Osler believes that his true calling is teaching. He has always been interested in how information is delivered and continues to explore the many different methods, models, and modes of instruction. His next endeavor in his academic career was the pursuit of a Master's degree in Educational Technology at his Alma Mater following his undergraduate studies. As a result of his graduate work, he began to explore and conduct investigations concerning Instructional Design and Technology. Techniques were developed that harnessed the power technology to enhance learning in the instructional setting. He seeks to bridge the gap between learning, instructional methods, and technological applications via collaborative and innovative problem–solving.

As Osler continued to develop his interests in Instructional Design and Technology, he pursued further graduate study at North Carolina State University. There he completed a Doctorate in Education with a research focus that culminated in his authoring the "Ergonomic Multimedia Courseware Equation ©". His current research focus involves the theories that explain the development and use of Technology and Ergonomic Metametric tools as effective learning methods.

This book is the culmination of his research efforts that have resulted in the invention of the "Trioengineering" Triological Science.

Currently, he is a Full Professor in the School of Education at NCCU. His interests include: a life filled with a love of Almighty GOD and service to his fellow man through: teaching; the production of innovative teaching tools; and groundbreaking research. He has been awarded three of the highest honors at NCCU as an employee and as faculty: The Employee Recognition Award for Outstanding Service in 2001, The University Award for Teaching Excellence in 2008, and the 1st Chancellor's Award for Innovation in 2014.

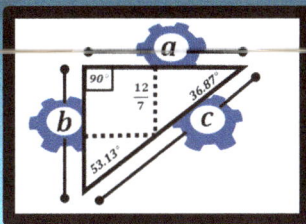

Give instruction to a wise man, and he will be yet wiser: teach a just man, and he will increase in learning.

Proverbs 9: 9

A Mathematical Guide:

This mathematical review is a detailed and comprehensive guide through mathematical concepts. It is designed to provide an enriching source of information. Mathematics is the foundation of all Visualus Calculations. This Guide can be viewed as both a starting point for initial study and as a reference tool that can be referred to at any time. Much of the information in the guide will be of value in the development of Visualus Equations. The Mathematical Guide begins with numerical definitions and concludes with a chart of mathematical symbols. Each section in this chapter builds upon the previous one so that the reader connects what they have learned with more detailed procedures and terminology. After reading this Chapter a foundation in mathematical concepts will be established that will greatly aid the reader as they begin to harness the power of inherent to Trioengineering.

A Detailed List of Mathematics Concepts and Techniques: Numbers Defined Mathematically for Trioengineering

"Natural Numbers" (also called "Counting Numbers")

1, 2, 3, 4, 5...

"Whole Numbers" ("Natural Numbers" including 0)

0, 1, 2, 3, 4, 5...

"Positive Integers" (also called "Natural Numbers")

1, 2, 3, 4, 5...

Common Mathematical Concepts Used in Equations

Exponent: a^2

An Exponent is a number that indicates repeated multiplication of a given number (referred to as the "Base Number"). Exponents are written as small numbers to the immediate upper right of the Base Number and indicate the repeated multiplication of the Base Number times itself. The Exponent number is referred to as the "Power" that the Base Number is being "Raised to" in mathematics.

A number raised to the second power is expressed as Base Number^2, and is called "Base Number Squared" or "the Base Number raised to the second power". "Multiple Powers" (greater than 2) can be illustrated in the same fashion as a Base Number that is raised to some designated power. For Example: a number raised to the third power is expressed as Base Number^3, and is called "Base Number Cubed".

Mathematical Examples of Exponents with Associated Properties:

Square Root

$x^2 = x \cdot x$

$3^2 = 3 \cdot 3 = 9$

Cubed Root

$x^3 = x \cdot x \cdot x$

$10^3 = 10 \cdot 10 \cdot 10 = 1000$

Large Root

$x^4 = x \cdot x \cdot x \cdot x$

$2^4 = 2 \cdot 2 \cdot 2 \cdot 2 = 16$

Properties of Exponents

Addition:

$$x^a x^b = x^{a+b}$$

A Mathematical Example of the Exponent Addition Property:

$$4^3 \cdot 4^5 = 4^{3+5} = 4^8$$

$$4^{3+5} = (4 \cdot 4 \cdot 4)(4 \cdot 4 \cdot 4 \cdot 4 \cdot 4) = 4^8$$

$$4^8 = 4 \cdot 4 \cdot 4 \cdot 4 \cdot 4 \cdot 4 \cdot 4 \cdot 4$$

Subtraction:

$$x^a / x^b = x^{a-b}$$

A Mathematical Example of the Exponent Subtraction Property:

$$4^6 / 4^2 = 4^{6-2} = 4^4$$

$$4^{6-2} = (4 \cdot 4 \cdot 4 \cdot 4 \cdot 4 \cdot 4)/(4 \cdot 4) = 4^4$$

$$4^4 = 4 \cdot 4 \cdot 4 \cdot 4$$

Multiplication:

$$\left(x^a\right)^b = x^{a \times b} = x^{ab}$$

A Mathematical Example of the Exponent Multiplication Property:

$$\left(4^2\right)^3 = 4^2 \cdot 4^2 \cdot 4^2$$

$$\left(4^2\right)^3 = (4 \cdot 4)(4 \cdot 4)(4 \cdot 4)$$

$$\left(4^2\right)^3 = 4^{2 \times 3} = 4^6$$

$$4^6 = 4 \cdot 4 \cdot 4 \cdot 4 \cdot 4 \cdot 4$$

Division:

$$(x/y)^a = x^a/y^a$$

A Mathematical Example of the Exponent Division Property:

$$(4/5)^2 = 4^2/5^2$$

$$(4/5)^2 = (4 \cdot 4)/(5 \cdot 5)$$

Exponential Product:

$$(x \cdot y)^a = (x \cdot y) \cdot (x \cdot y) \cdot (x \cdot y)... = x^a \cdot y^a$$

A Mathematical Example of the Exponential Product Property:

$$(4 \cdot 5)^2 = (4 \cdot 5)(4 \cdot 5) = 4^2 \cdot 5^2$$

Negative Exponents

$$x^{-a} = 1/x^a = x^{-1} = 1/x$$

A Mathematical Example of a Negative Exponent:

$$4^{-2} = 1/4^2 = 1/(4 \cdot 4)$$

Exponent = Zero:

$x^0 = 1$, for all x where, $(x \neq 0)$

A Mathematical Example of Zero as an Exponent:

$$4^0 = 1$$

Exponent = One:

$x^1 = x$, for all x

A Mathematical Example of One as an Exponent:

$4^1 = 4$

Fractional Exponents:

Fractional Exponents are the same as taking a root. Thus, $x^{\frac{1}{2}} = \sqrt{x}$ this is true if we examine the Exponential Multiplication Property.

Mathematical Examples of Fractional Exponents:

$(x^{\frac{1}{2}})^2 = x^{\frac{1}{2} \times 2} = x^{\frac{1}{2} \times 2/1} = x^{2/2} = x^1 = x$, Thus,

$x^{1/a} = \sqrt[a]{x}$, and

$x^{a/b} = (\sqrt[a]{x})^b$, Example:

$4^{\frac{1}{2}} = \sqrt{4} = 2$, because $2 \cdot 2 = 4$

$$4^{\frac{1}{2}} = \left(\sqrt{4}\right)^{1} = (2)^{1} = 2^{1} = 2$$

$$64^{2/3} = \left(\sqrt[3]{64}\right)^{2} = (4)^{2} = 4^{2} = 4 \cdot 4$$

Radical

The "Radical" is a symbol used in determining the "Root" of a given number. The mathematical symbol for a Radical is: " $\sqrt{}$ ". The small number in front of the Radical is called the "Index" of the Radical. The Index indicates that the number in the Index is the Root to be calculated. A traditional Square Root has an Index of 2, however, this is generally not expressed because a Root of 2 is the basic starting point for all Root calculations. Thus, a Root of 2 (referred to in mathematics as a "Square Root") simply uses a plain Radical symbol with a number. Root equations higher than 2 will use the number in the Index of the Radical.

Mathematical Examples of the Use of the Radical Symbol:

$\sqrt[3]{x}$ = The "Cubed Root" of x, because $\left(\sqrt[3]{x}\right)^{3} = x$

Thus, $\sqrt[3]{8}$ = 2, because 2 · 2 · 2 = 8

Other Examples:

$\sqrt[3]{8}$ = 2, because 2 · 2 · 2 = 27

$\sqrt[5]{32}$ = 2, because 2 · 2 · 2 · 2 · 2 = 32

$\sqrt[4]{10,000}$ = 10, because 10 · 10 · 10 · 10 = 10,000

Root

The Root of an equation is also the solution to that particular equation literally meaning a number times itself will produce a given number. Taking the Root of a number is the opposite of raising a number to a power (or raising a number by an exponent). A Root uses the " $\sqrt{}$ " symbol also known as the "Radical Symbol". The Square Root of a number is expressed as: $\sqrt{x}$. This is the opposite of raising x to the second power or $x^{1/2} = \left(\sqrt{x}\right)^1 = \sqrt{x} = x$. Roots can be expressed as fractions. Thus, "$x^{1/2}$" illustrated previously.

A Mathematical Example of a Root:

$$\sqrt[a]{x} = x^{1/a}$$

$$\sqrt[5]{32} = 32^{1/5}$$

Square Root

Square Roots obey the property that $\sqrt{ab} = \sqrt{a}\sqrt{b}$.

A Mathematical Example of a Square Root:

$$\sqrt{225} = \sqrt{9 \cdot 25} = \sqrt{9} \cdot \sqrt{25} = 3 \cdot 5 = 15$$

Because, $15 \cdot 15 = 225$

Thus, $\sqrt{225} = 15$

The Square Root of a number is written as: $\sqrt{x}$. This is also known as "x Raised to the One–Half Power": $x^{\frac{1}{2}} = \left(\sqrt{x}\right)^1 = \sqrt{x} = x$.

Further Examples of Square Root:

$\sqrt{1} = 1, \sqrt{4} = 2, \sqrt{9} = 3, \sqrt{16} = 4, \sqrt{25} = 5, \sqrt{36} = 6...$

This is true because,

$1 \cdot 1 = 1, 2 \cdot 2 = 4, 3 \cdot 3 = 9, 4 \cdot 4 = 16, 5 \cdot 5 = 25, 6 \cdot 6 = 36...$

The Square Root of most Integers will be an Irrational Number.

A Mathematical Example of an Irrational Square Root:

$\sqrt{2} = 1.414...$

Squares and Square Root

A Mathematical Example of the Exponential Meaning of Square Root is "x to the Power of 2" or "x Squared" or "The Square Root x" written as:

$x^2 = x \cdot x$

$3^2 = 3 \cdot 3 = 9$

Cubed Root

A Mathematical Example of the Exponential Meaning of Cubed Root is

"x to the Power of 3" or "x Cubed" or "The Cubed Root of x" written as:

$x^3 = x \cdot x \cdot x$

$10^3 = 10 \cdot 10 \cdot 10 = 1000$

Absolute Value

The Absolute Value is a non–negative number equal in numerical value to a given real number. The Absolute Value is always given to a Real Number. An Absolute Value is always positive and/or zero. Remember: All Real Numbers are represented on a number line; the Absolute Value can be considered the distance the number is from zero (distances cannot be negative or measured by negative numbers). Absolute Value of a number is symbolized by $|x| = x =$ "The Absolute Value of x". An excellent example of Absolute Value is distance traveled from point A to point B. The distance traveled from point A to arrive at point B is a positive real number that can represented by the Absolute Value of that number which is the amount of distance traveled by some measurable value (such as miles or kilometers). The distance traveled from point A to point B cannot be represented by a negative number (as one does not measure or travel a distance negatively). Mathematical Examples of Absolute Value:

$$|x| = x$$

$$|17| = 17$$

$$|3.5| = 3.5$$

$$|-10| = 10$$

$$|4| = 4$$

$$|0| = 0$$

Mathematical Operations for Trioengineering

Variable

A Variable is a quantity (sometimes unknown) that may assume any number from a set of values. In Algebra, a variable is a letter used to represent values from the set of Real Numbers. A variable is also a quantity that can change or that may take on different values. The term "Variable" is also used to directly refer to a letter or a symbol representing an unknown quantity in an equation.

Algebra

Algebra is the study of the properties of operations carried out on sets of numbers. Algebra is a branch of mathematics in which symbols, usually letters of the alphabet are used to indicate a given or an unknown number.

Operations

An operation is the process of carrying out a rule on a set of numbers. The 4 fundamental mathematical arithmetic operations are: Addition, Multiplication, Subtraction and Division. Each Operation is listed with its associated properties in the following manner:

1. Addition

Addition is the operation of combining two numbers to form a sum. An example of addition is 3 + 4 = 7.

The Operation of Addition has four Properties:

A. The Commutative Property

$a + b = b + a$, for all a and b

Thus, 3 + 4 = 4 + 3

B. The Associative Property

$(a + b) + c = a + (b + c)$, for all a, b, and c

Thus, (3 + 4) + 2 = 3 + (4 + 2)

C. The Additive Identity Property

$a + 0 = a$, for all a

Thus, $4 + 0 = 4$

D. The Distributive Property

$a \times (b + c) = a(b + c) = a(b) + a(c)$, for all a, b, and c

Thus, $4 \times (3 + 2) = 4(3 + 2) = 4(3) + 4(2)$

2. Multiplication

Multiplication is the operation repeated of the Addition. Examples of various equations that symbolize Multiplication are as follows: $\times$, $\cdot$, and $*$. Thus, $3 \times 5 = 3 \cdot 5 = 3 * 5 = 3*5 = 3(5) = 15$.

The Operation of Multiplication results in an equation that yields a total number called "the Product". The equation that produces the Product combines several numbers called "Multiplier Groups" of similar size referred to as the "Multiplicand". If we combine 3 groups with 4 objects in each set, the same answer can be obtained by the operation of addition. In the aforementioned 3 group 4 set example, the Operation of Addition equation would be: $4 + 4 + 4 = 12$; which is the equivalent of the Operation of Multiplication equation: $3 \cdot 4 = 12$. Thus, the Operation of Multiplication is in fact, repeated Addition.

In Algebra, symbols are often removed from the Operation of Multiplication and numbers are replaced by letters called "variables". An example of this are as follows: ab, πr^2, $\frac{1}{2}ar^2$ in which each respectively means the following: $ab = a \cdot b$, $\pi r^2 = \pi \cdot r^2$, and $\frac{1}{2}ar^2 = \frac{1}{2} \cdot (a \cdot r^2)$.

Equations

An equation is a mathematical statement that linearly expresses the equality of two mathematical expressions that have the same value. The equal symbol "=" is used to indicate equality on both sides of the equation.

Examples:

$4 \cdot 5 = 20$

$10^2 = 10 \cdot 10 = 100$

$\frac{1}{2} = 1/2 = 0.5$

$10 + 25 = 35 = 5 \times 7$

Function

A function is a rule that turns each member of one set into a member of another set. The most common functions are functions that turn one number into another number. Example: The function ƒ(x) (pronounced "ƒ of x") can be written as the following: ƒ(x) = 3x2 + 5 can be applied to the set: x = {1, 2, 3}. Thus, the new set becomes {8, 17, 32}...

Because,

For $x = 1$, $f(x) = 3(1)^2 + 5 = 8$

For $x = 2$, $f(x) = 3(2)^2 + 5 = 17$

For $x = 3$, $f(x) = 3(3)^2 + 5 = 32$

A function is also a relation that uniquely associates members of one set with members of another set. A function is therefore a "many–to–one" (or sometimes a "one–to–one") relation. The set A of values at which a function is defined is called its "Domain", while the set B of values that the function can produce is called its "Range".

Thus, a Function changes one number into another number by inputting a value into the Domain and outing it as new value in the Range.

Further Examples of Functions:

Function: $y = 2x^2$

Where,

y = Domain, 2x = Range

Thus, the Function y = $2x^2$, changes the values in the set {1, 2, 3} when placed in the Range: {$2(1)^2$, $2(2)^2$, $2(3)^2$} to the Domain: {y = 2, y = 8, y = 18} = {2, 8, 18}.

Function: $f(x)$ = 2x + 1

Where,

$f(x)$ = Domain, 2x + 1 = Range

Thus, the Function $f(x)$ = 2x + 1, changes the values in the set {1, 2, 3} when placed in the Range: $f(1)$, $f(2)$, $f(3)$ to the Domain: $f(1)$ = 2(1)+1, $f(x)$ = 2(2)+1, $f(3)$ = 2(3)+1, thus the domain set = {3, 5, 7}.

Binomial Expansion

A Binomial is a Polynomial with two terms. The Binomial Expansion Theorem is a shortcut method of raising a Binomial to a power. Binomial Expansion raises equations by a power. Examples of Binomial Expansion are as follows:

$$(x + y)^0 = 1$$

$$(x + y)^1 = x + y$$

$$(x + y)^2 = (x + y)(x + y) = x^2 + 2xy + y^2$$

$$(x + y)^3 = (x + y)(x + y)(x + y) = x^3 + 3x^2y + 3xy^2 + y^3$$

$$(x + y)^4 = (x + y)(x + y)(x + y)(x + y) = x^4 + 4x^3y + 6x^2y^2 + 4xy^3 + y^4$$

$$(x + y)^5 = (x + y)(x + y)(x + y)\ (x + y)(x + y) =$$

$$x^5 + 5x^4y + 10x^3y^2 + 10x^2y^3 + 5xy^4 + y^5$$

Qualities of the Binomial Expansion

The following qualities apply to Binomial Expansion:

1.   There are $(n + 1)$ terms in the expansion of $(x + y)^n$.

2.   The degree to each term is n.

3.   The powers on "x" begin with n and decrease to 0.

4.   The powers on "y" begin with 0 and increase to n.

5.   The coefficients are symmetric.

The Binomial Expansion Theorem

The Binomial Expansion Theorem explains in detail the process of Binomial Expansion. Written in Sigma or Summation Notation the Binomial Expansion Theorem is as follows:

$$(x+y) = \sum_{k=0}^{n} \binom{n}{k} x^{n-k} y^k$$

In its expanded form the Binomial Expansion Theorem is:

$$(x+y)^n = \binom{n}{0}x^n + \binom{n}{1}x^{n-1}y + \ldots + \binom{n}{k}x^{n-k}y^k + \ldots + \binom{n}{n-1}xy^{n-1} + \binom{n}{n}y^n$$

Factorials

The factorial of a number is the product of all the whole numbers, except zero, that are less than or equal to that same number. The factorial of a number is expressed by placing an exclamation point after that number. Thus, "the factorial of 7" is written as: "7!". The factorial of 7 is expressed in an equation as $7! = 7 \cdot 6 \cdot 5 \cdot 4 \cdot 3 \cdot 2 \cdot 1 = 5040$. The factorials 1! – 12! are as follows:

$1! = 1 = 1$

$2! = 2 \cdot 1 = 2$

$3! = 3 \cdot 2 \cdot 1 = 6$

$4! = 4 \cdot 3 \cdot 2 \cdot 1 = 24$

$5! = 5 \cdot 4 \cdot 3 \cdot 2 \cdot 1 = 120$

6! = 6 · 5 · 4 · 3 · 2 · 1 = 720

7! = 7 · 6 · 5 · 4 · 3 · 2 · 1 = 5040

8! = 8 · 7 · 6 · 5 · 4 · 3 · 2 · 1 = 40,320

9! = 9 · 8 · 7 · 6 · 5 · 4 · 3 · 2 · 1 = 362,880

10! = 10 · 9 · 8 · 7 · 6 · 5 · 4 · 3 · 2 · 1 = 3,628,800

11! = 11 · 10 · 9 · 8 · 7 · 6 · 5 · 4 · 3 · 2 · 1 = 39,916,800

12! = 12 · 11 · 10 · 9 · 8 · 7 · 6 · 5 · 4 · 3 · 2 · 1 = 479,001,600

Factorials are useful because they show all of the possible numerical arrangements that exist to a given set of things. For example, if 5 books are lined up side by side on a shelf and we want to know every possible way to arrange them we can simply take the factorial of 5. Thus, 5! = 5 · 4 · 3 · 2 · 1 = 120. This illustrates that we can arrange the set of all 5 books into 120 different orders until we have exhausted all of the possible ways to arrange them. It is also useful to note that in mathematics the factorial of "0!" is equal to the set of 1. This is true because the arrangement of an empty set is still a single arrangement thus, 0! = 1.

Explaining the **Combination Equation** and **Binomial Coefficient:** The Combination Equation is a mathematical equation that uses the Binomial Coefficient and Factorials to determine the total number of arrangements of a set of objects without repetition. The value of the Combination Equation is that it also gives the total number of arrangements that are possible. The Combination Equation is as follows:

$$_nC_r = \binom{n}{r} = \frac{n!}{(n-r)r!}$$

Where,

n = The total number of objects to choose from;

r = The number of objects in the arrangement; and the equation includes the following:

The Binomial Coefficient

$\binom{n}{r}$ of the Combination Equation is called the Binomial Coefficient. It represents the rth element in the nth row of Pascal's Triangle.

Example of the Use of the Combination Equation

If an instructor gives 10 questions on a quiz in class today, and states the next test will consist of 5 questions out of those 10. If the order does not matter, then how many different tests can be created?

$$\text{Number of Tests} = \binom{10}{5} = \frac{n!}{(n-r)r!}$$

$$= \binom{10}{5} = \frac{10!}{(10-5)5!} = \frac{10!}{(5!)5!} = \frac{10 \cdot 9 \cdot 8 \cdot 7 \cdot 6 \cdot 5 \cdot 4 \cdot 3 \cdot 2 \cdot 1}{(5 \cdot 4 \cdot 3 \cdot 2 \cdot 1)5 \cdot 4 \cdot 3 \cdot 2 \cdot 1} =$$

$$\left[\text{Factor Out } (5 \cdot 4 \cdot 3 \cdot 2 \cdot 1)\right] = \frac{10 \cdot 9 \cdot 8 \cdot 7 \cdot 6}{5 \cdot 4 \cdot 3 \cdot 2 \cdot 1} = \frac{30240}{120} = 252$$

Thus, a total 252 different tests are possible.

Logarithm: The Natural Logarithm

A Natural Logarithm is a logarithm in which the base is the Irrational Constant Number approximately equal to "e" = 2.718281828 . . . Symbol "$\ln(x)$" or "$\log_e (x)$" or "$\log(x)$" with the base "e" implicit. The Natural Logarithm of a given number "x" is the power to which the base "e" would have to be raised to be equal to x. Where, the rule for Logarithms is $x = b^y$, because $y = \log_b(x)$, thus, $\log_e 10 \approx 2.30258$ because for e = 10, the number e must be raised to 2.30258 or $e^{2.30258} = 9.9999\ldots \approx 10$. The Natural Logarithm is also sometimes called the "Napierian Logarithm".

The Independent Variable

The Mathematical definition of the Independent Variable is parallel to the statistical definition that is the Independent Variable = Treatment. You can see the obvious similarities in the following definition: "In Mathematics an Independent Variable is a Variable in an equation that may have its value freely chosen without considering values of any other variable. For equations such as y = 3x − 2, the independent variable is x. The variable y (dependent variable or value) is not independent since it depends on the number chosen for x." Formally, an independent variable is a variable which can be assigned any permissible value without any restriction imposed by any other variable.

Mathematical Independent Variable Examples:

Equation	Independent Variable(s)	Dependent Variable
$y = x^2 - 2x + 3$	x	y
$x^2 + y^2 = 1$	x	y
$x = 1 - t^2$	t	x
$z = 2x^2 - y^3$	x and y	z

A list of Mathematical Equation Symbols and their relative meanings is presented for Visualus Equation reference on the pages that follow.

Equation Symbols and Their Respective Meanings

Symbol	Mathematical Definition	Mathematical Meaning	Expression Definition	Example
( )	Parentheses	"Quantity"	Denotes A Quantity	$(x + y)$
[ ]	Square Brackets	"The Quantity"	Denotes A Quantity	$w + [(x + y) + z]$
=	Equal Sign	"Equals"	Indicates Two Values Are The Same	$-(-5) = 5$
≈	Approximate Equal Sign	"Is Approximately Equal To"	Indicates Two Values Are Close To Each Other	$x + y \approx z$
≤	Inequality Sign	"Is Not Equal To"	Indicates Two Values Are Different	$3 \leq 5$
<	Inequality Sign	"Is Less Than"	Indicates Value On Left Is Smaller Than Value On Right	$3 < 5$
≥	Inequality Sign	"Is Less Than Or Equal To" "Is At Most Equal To"	Indicates Value On Left Is Smaller Than Or Equal To Value On Right	$x \geq y$
>	Inequality Sign	"Is Greater Than"	Indicates Value On Left Is Larger Than Value On Right	$5 > 3$
≠	Inequality Sign	"Is Greater Than Or Equal To"	Indicates Value On Left Is Larger Than Or Equal To Value On Right	$x \neq y$
| |	Absolute Value Sign	"The Absolute Value of"	Distance Of Value From Origin In Number Line, Plane, Or Space	$\|-3\| = 3$

Equation Symbols and Their Respective Meanings

Symbol	Mathematical Definition	Mathematical Meaning	Expression Definition	Example
+	Addition Sign	"Plus"	Sum of Values	$3 + 5 = 8$
*	Multiplication Sign	"Times"	Product Of Two Values	$A * B = B * A$
x	Multiplication Sign	"Times"	Product of Two Values	$3 \times 5 = 15$
•	Multiplication Sign	"Times"	Product of Two Values	$3 • 5 = 3 \times 5 = 15$
−	Subtraction Sign Minus Sign	"Minus" or "Negative"	Difference of Two Values, Negative Number	$3 - 5 = -2$
±	Plus/Minus Sign	"Plus" Or "Minus"	Expression Of Range	$500 ± 10\%$
^	Carat	"To The Power of"	Exponent	$2\text{^}5 = 2^5 = 32$
!	Exclamation	"Factorial"	Product Of All Positive Integers Up To A Certain Value	$5! = 5 \times 4 \times 3 \times 2 \times 1 = 120$
√	Surd or Square Root Symbol	"The Root of" or "The Square Root of"	Algebraic Expressions	$\sqrt{4} = 2$
...	Continuation Sign	"And So On Up To" "And So On Indefinitely"	Extension Of Sequence	$S = \{1, 2, 3, ...\}$
/	Slash	"Divided By" "Over"	Division	$3/4 = 0.75$
÷	Division Sign	"Divided By"	Division	$3 ÷ 4 = 0.75$
%	Percent Symbol	"Percent"	Proportion	$0.032 = 3.2\%$
:	Ratio Sign	"Is To" "Such That" "It Is True That"	Division Or Ratio, Symbol Following Logical Quantifier Or Used In Defining A Set	$1:2 = 10:20$
∞	Lemniscate	"Infinity" "Increases Without Limit"	Infinite Summations Infinite Sequence Limit	$x < ∞$

Chapter Two follows provides the foundations of Trioengineering.

TRIOENGINEERING ™ © *The Problem-Solving Triological Science: The In-Depth Trichotomous Science of the Dynamic 3-4-5-6 Golden Upright Right Triangle for Innovative Problem-Solving.* Osler Studios Incorporated ©, © Copyright 2022 All Rights Reserved.

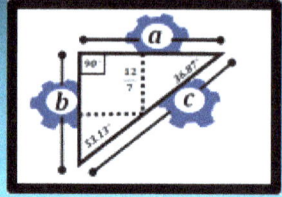

Every good gift and every perfect gift is from above, and cometh down from the Father of lights, with whom is no variableness, neither shadow of turning.

James 1: 17

Trioengineering Terminology & Concepts

The "∇" [The "Trichotomous Upright Right Triangle"] as the Foundation of All of the Triological Sciences

The Triological Sciences (which are identified as—(a.) Triology; (b.) Trithmetic; (c.) Triomathematics; (d.) Triophysics; (e.) Triostatistics; (f.) Trioengineering; (g.) Trichometry; and (h.) Triangulus) are all grounded by the same foundation that has its roots in the geometric "Golden Upright Right Triangle" also known as the "3-4-5-6 Golden Upright Right Triangle" (represented by the acronym "GURT"). The Golden Upright Right Triangle is derived as an internal characteristic from the "Visualus Isometric Cuboid © ™" and is symbolized by the "Trine abc" = "∇abc". The Trine symbol is also the mathematical symbol for the "Trichotomous Upright Right Triangle" as the triple terms "Trichotomous Upright Right Triangle" as [∇ = "The Trichotomous Upright Right Triangle" or "Upright Right Trine"]. Therefore, the appearance of the Trine symbol before a title in the Triological Sciences indicates a shorthand notation for the words "Trichotomous Upright Right Triangle" directly referring to the

"Golden Upright Right Triangle" (as such "Trine *abc*" = "∇abc" = "Golden Upright Right Triangle of *abc*"). The true foundation of all of the "Triological Sciences" is the base underpinning of mathematical model of Visualus which is the Isometric Cuboid. The Visualus Isometric Cuboid is presented graphically below:

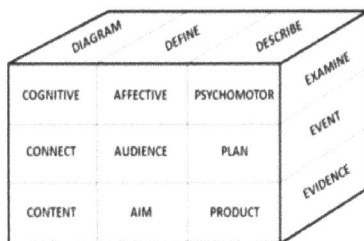

"Visualus" is the mathematics of innovative, inventive, and in-depth problem-solving via the geometric representation of the "Rectilinear Model of Instructional Systems Design". The next series of models and equations with explain in explicit detail and meticulous precision exactly how ∇abc is derived from the Visualus Isometric Cuboid as the inherent foundation and true beginning of all of the Triological Sciences. The Trichotomous Upright Right Triangle of *abc* is defined in Cartesian Coordinates mathematically in the following "Golden Upright Right Triangle Triological Side to Coordinate Mathematical Definition Equation":

$$\nabla abc = \nabla[abc] = \nabla[xyz] = \nabla[x] \cdot \nabla[y] \cdot \nabla[z] =$$
$$\nabla[a] \cdot \nabla[b] \cdot \nabla[c] \text{ for } \nabla[\text{Side } a] \cdot \nabla[\text{Side } b] \cdot \nabla[\text{Side } c] = \nabla abc.$$

According to the above " ∇Triological Side to Coordinate Mathematical Definition Equation", the following applies:

The holistic all-inclusive "GURT" is represented by "∇" = *Trioengineering* = Trine;

The measurable unit "Length" is represented by "x" = x–coordinate = abscissa;

The measurable unit "Height" is represented by "y" = y–coordinate = ordinate;

The measurable unit "Depth" is represented by "z" = z–coordinate = applicate; and

The mathematical operator "[ ]" is equitable to "a concentration on" = "to focus on".

Graphically, the relationship between ∇abc and the Visualus Isometric Cuboid is illustrated as follows:

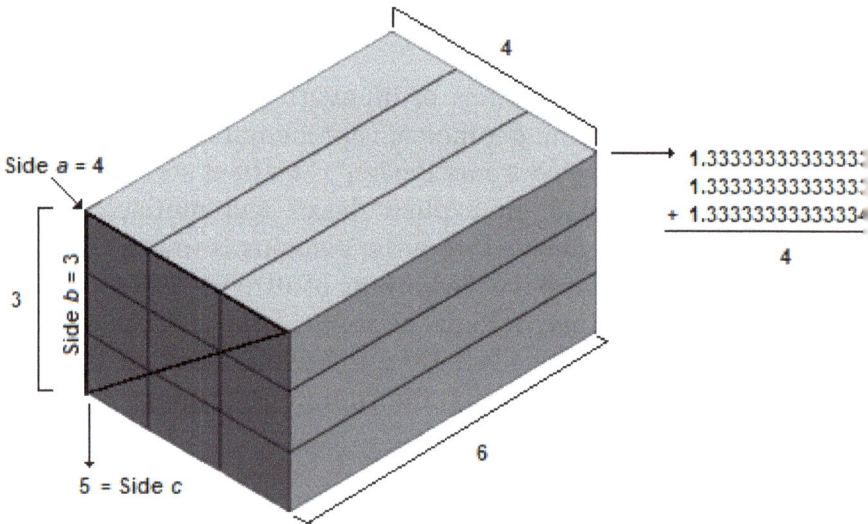

Thus, from ∇abc which is extracted out of the Visualus Isometric Cuboid: $[xyz]$ = $[x][y][z]$ illustrating the three Cartesian Coordinates. Idealistically, in terms of vectors, if Cartesian Coordinates x and z are respectively perpendicular then their cross product is Cartesian Coordinate y. The reduction of Cartesian Coordinate z to present only the Front Face of the Visualus Isometric Cuboid that is also the 3 by 3 Standard Table Format of the Triostatistics Tri–Squared Test. The mathematics of the removal of Cartesian Coordinate *z* to have only the Front Face of the Isometric Cuboid as the 3 by 3 Table remain is illustrated mathematically and graphically in the sections that follow.

Defining Triological Science Notation: "Total Trioengineering Notation" or "*Trioengineering Notation*"

"Trioengineering Notation" is used to mathematically parsimoniously explain the "mathematical law of trichotomy" through the use of 3-4-5-6 Golden Upright Right Triangle (or "GURT") and its properties as a holistic mathematical operation in the same manner as Summation or "Sigma" Notation (that uses "Σ") and Product Notation (that uses "Π"). Trioengineering Notation has a variety of uses and applications in the various Tiological Sciences. The symbol for all Trioengineering Notation is the "Triune" or "∇" (that when used in Trioengineering Notation literally means "Trioengineered" which is a shorthand way of saying "The Trioengingeering 3-4-5-6 Golden Upright Right Triangle"). Whenever and wherever the Trioengineered Trioengineering Notation Triune "∇" symbol is used it indicates that the equation, model, solution, and/or calculation uses and is a part of Trioengineering as a whole via the 3-4-5-6 Golden Upright Right Triangle. As the GURT is the universal and connective foundation of all of the Triological Sciences and all of Trichotomous Research—Trioengineering Notation is therefore universally used).The notation has broad utility and used in multiple ways using the following formats that are detailed in all of the sections and equations, models, solutions, and calculations that follows.

Trioengineering Notation has the same nomenclature parameters as the more traditional "Summation Notation" that uses the upper case Greek letter "Sigma" ("Σ") to indicate summation in a series and "Product Notation" that uses the upper case Greek letter "Pi" ("Π") to indicate multiplication in a series. Trioengineering Notation is used in this format connote geometric trichotomous equations, calculations, and formulae that directly pertain to the utility and viability of the 3-4-

5-6 Golden Upright Right Triangle in problem-solving and the Triological Sciences in particular. In this format, Trioengineering Notation is written as follows to connote the triple trichotomy of a concept, idea, thought and/or solution (that corresponds directly to the triune trichotomy of the three angles and three sides of the GURT) as follows:

$$\overset{3}{\underset{i=1}{\bigtriangledown}}$$

Where,

$\bigtriangledown$ = The "**Triune**" Symbol that literally means "**Trioengineered**" or "**Trioengineering**" = "The Trichotomous Upright Right Triangle" = "The 3-4-5-6 Golden Upright Right Triangle" (or "GURT");
3 = The immutable unchanging ending point called a "**Triand**" for "Trichotomous Combination End"; and
1 = The starting point called a "**Triart**" for "Trichotomous Start" (that always at $\angle \alpha$ and/or Side a depending upon the model, for example Triostatistics Triangular Equation Modeling or [TEM] that always starts with "Side a" versus the cyclical Trioengineering Model of ["author"; "build"; and "convey"] that always starts with "$\angle \alpha$") on the $\bigtriangledown$ even when the model is cyclical and ongoing; and
i = The index (or the "Starting Point") that is always at 1 on the $\bigtriangledown$.

Trioengineering Notation is used in geometrical names to connote the actual application of the "3-4-5-6 Golden Upright Right Triangle" in its entirety, as in this example:

" $\bigtriangledown abc$" (that can be literally defined and referred to as "Triune abc" or "Trioengineered abc").

It is also used in mathematical formulas and equations as illustrated in this example:

$$\nabla[\ \nabla y = \nabla mx + \nabla b].$$

Used as the preconditional modifier that explains the application of the Golden Upright Right Triangle in defining in-depth and precise mathematical operations as GURT conditions (using both or either "Tri" and " ∇") as exhibited below:

$$\underset{name}{\overset{n=3}{\mathrm{Tri}}} = \underset{name}{\overset{Tri}{\nabla}}\ ; and$$

$$\underset{name}{\overset{n=3}{\mathrm{Tri}}}[calculation\ and/or\ formula] = \underset{name}{\overset{Tri}{\nabla}}[calculation\ and/or\ formula].$$

Additionally, Trioengineering Notation is used to convey the Tripositive nature of all positive integers as defined in explicit detail in the next section. Trioengineering format when used to define all positive numbers (including zero) in this manner is written and expressed as follows:

$$\overset{n=3}{\underset{i=1}{\underset{num}{\nabla}}}.$$

The Final Full and Complete Mathematical Definition of Trioengineering Notation

Trioengineering Notation is fully and completely mathematically defined in the following definitive mathematical series of equations presented below.

Trioengineering is meticulously, precisely, specifically used to parsimoniously define the trichotomous use of the 3-4-5-6 Upright Right Triangle in the following mathematical equations:

$$V = \overset{3}{\underset{i=1}{V}} = \text{Vertical Abbreviation } V \text{ (with notation on the Trine)} = \text{The}$$

definition of all Positive Integers (including Zero because it completely satisfies the definition and conditions of Upright Right Triangle numbers) Upright Triangle Number applicable to the measurements = V Name or Abbreviation of Measure = [Tri] as the "Tripositive Trichotomy" where, [V= "Tri"] = The Geometric Models as the ""CTCG Function"" = where CTCG = ["Convert" to by "Transforming" into and "Conforming" to "Gain"] = The "3-4-5-6 Golden Upright Right Triangle" that is also referred to as the "Trichotomous Upright Right Triangle" geometrically to =

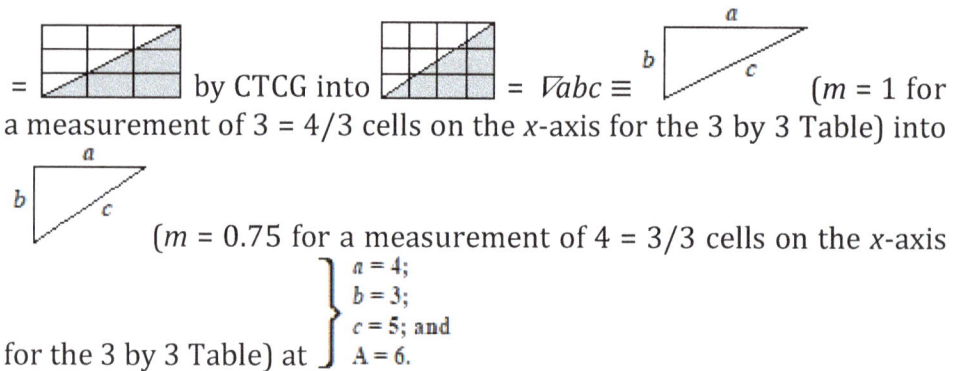

= by CTCG into = $\sqrt{abc} \equiv$ (m = 1 for a measurement of 3 = 4/3 cells on the x-axis for the 3 by 3 Table) into

 (m = 0.75 for a measurement of 4 = 3/3 cells on the x-axis

for the 3 by 3 Table) at $\left.\begin{array}{l} a = 4; \\ b = 3; \\ c = 5; \text{ and} \\ A = 6. \end{array}\right\}$

Triological Science Triomathematics: Trichotomous Upright Right Triangle Numbers

Johann Carl Friedrich Gauss was a noted and well known 18th Century German mathematician and physicist who made significant contributions to many fields in both math and science. He is universally considered to be one of history's most influential mathematicians. In Gauss's diary entry on July 10th, 1796 he made a great discovery related to sum of triangular numbers that stated: "**EYPHKA! num = Δ + Δ + Δ**" and in addition records his discovery of a proof that any number is the sum of 3 triangular numbers. This is more commonly known as Gauss's Eureka Theorem. The Theorem is modified by the author into the "Trichotomous Upright Right Triangle" Theorem and uses Trioengineering Notation to rewrite the original theorem in the context of the Trichotomous Upright Right Triangle into the following:

$$\sum_{i=1}^{n=3} \nabla = \nabla_1 + \nabla_2 + \nabla_3 \ .$$

The above "Trioengineering Notation Trichotomous Upright Right Triangle Equation" defines all positive numbers as true "Tripositive Trichotomous Upright Right Triangle Integers" (including the number "0"). As such, all "Tripositive Trichotomous Upright Right Triangle Integers" have the following trichotomous properties: (1.) "Magnitude" (as "Size"); (2.) "Distance" (as "Length"); and (3.) "Position" (as "Place") in a self-contained inharmonic summative "Triharmonic Triune". These numeric properties are valid characteristics that are reflected in nature as "Triologically Scientifically" viable and tangible with an intentional sub-atomic infrastructure that is the initial structural framework for all of existence, all of reality, and all of nature.

All positive integers including zero "0" are "Trichotomous Upright Right Triangle Numbers" and their respective values can all be trichotomously created in a trifold manner via a threefold Tripositive calculation. The Tripositive calculation is satisfied by the equation: $[(n^2 + n)/2]$, which is the mathematical formula used to create the Triangular Numbers need to complete all positive integers including zero as "Trichotomous Upright Right Triangle Numbers". The Tripositive is... The Tripositive is written in the following manner:

"[value]". In the case of the "Trichotomous Upright Right Triangle Numbers", the Tripositive number is written as "[num]" and is parsimoniously rewritten as, "[n]", thus, "[num] = [n]", for all positive integers including zero ("0").

The Golden Upright Right Triangle Vector Model displaying how the 3-4-5-6 Golden Upright Right Triangle positing conveys the Pythagorean Theorem to Produce the Acclivity of Side *c*:

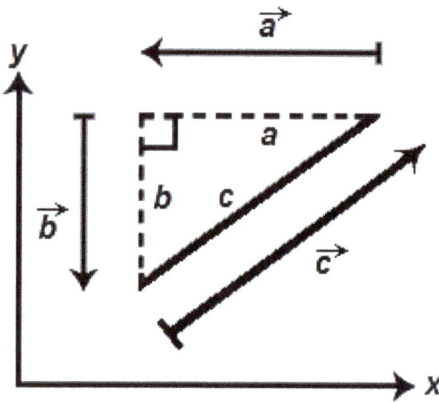

Where,

a = x

b = y

For all Unit Values as Vectors that become the 3-4-5-6 GURT Sides respectively.

Note:

If $a = \begin{bmatrix} x \\ y \end{bmatrix}$ then the size of $|c| = \sqrt{(x^2 + y^2)}$

Side *c* can be mathematically expressed as a unit value in the 3-4-5-6 Golden Upright Right Triangle as the "Total Side *c* [3-4-5-6 GURT] Equation" as a mathematical formula:

$$\text{Side } c = \text{Side } a \left[\sqrt{1 + \left(\frac{\text{Side } b}{\text{Side } a} \right)^2} \right].$$

The aforementioned can be further simplified into a more "Complete Form" of the "Total Side c Equation" that is written in the following manner:

$$c = a\left[\sqrt{1 + \left(\frac{b}{a}\right)^2}\right].$$

Upon the input of the three respective 3-4-5-6 GURT Side Values as Side $a = 4$; Side $b = 3$; the final "Calculated Form" of the "Total Side c Equation" will yield Side $c = 5$ as illustrated below:

$$5 = 4\left[\sqrt{1 + \left(\frac{3}{4}\right)^2}\right].$$

According to the abovementioned 3-4-5-6 GURT Vector Model, "the acclivity" (which literally means "the upward slope" of Side c = Side b over Side a) of side c is mathematically defined as $b/a = \frac{3}{4} = 0.75$. In the 3-4-5-6 Golden Upright Right Triangle Side $a = 4$; Side $b = 3$; and Side $c = 5$ (therefore as illustrated in the model above the respective x and y coordinate abscissa and ordinate as vectors {a line with both direction and magnitude} measures out to x = Side a = 4 Unit Side Vector = $\vec{a}$, with y = Side b = 3 as the Unit Side Vector = $\vec{b}$, and [x, y] = Side c = 5 as the Unit Side Vector = $\vec{c}$, respectively, therefore, the equation, $|c| = \sqrt{(x^2 + y^2)}$ is satisfied because, $|5| = \sqrt{(4^2 + 3^2)} = \sqrt{(16 + 9)} = \sqrt{25} = 5$); with an overall area of A = 6. The 3-4-5-6 Golden Upright Right Triangle is the basis for the Triological Science Triostatistics "Triangular Modeling Equation also known as "[TEM]". [TEM] is used to model a vast variety of trichotomous relations and relationships between ideas, concepts, and solutions in research. The 3-4-5-6 Golden Upright Right Triangle characteristics are as follows: .

Triangle Numbers Expressed via the Total Triangle Notation that is "Trioengineering Notation"

The Golden Upright Right Triangle Identity for all Positive Integers (including Zero) written in "Upright Right Total Triangle Notation" in following manner:

$$\sum_{i=1}^{n=3} \nabla_{num} \equiv \nabla_1 + \nabla_2 + \nabla_3 .$$

The solution above is based on mathematician Carl Gauss' solution from his 1796 notebook which stated the following as the first presentation of the Triangle Number Theorem as a Triangular Number proof which originally was entered on 07.10.1796 and read as follows (note: "EYPHKA!" written in Greek which is translated into English as: "Eureka!") was originally written as follows:

EYPHKA! num = Δ + Δ + Δ ,

The "Mathematical Identity for Upright Right Triangle Numbers" (which provides each Upright Right Triangle Number with a mathematical definition that is equivalent to Gauss' original Triangle Number Proof. An "Upright Right Triangle Number" (for all positive integers is represented by the following notation and symbol: "$\sum_{i=1}^{n=3} \nabla_{num}$".

The aforementioned is the notation for all positive integers (including zero) that are composed of the sum of three Upright Right Triangle Numbers. The symbol " ∇" (an italicized "nabla") is used in this case to indicate the "Golden Upright Right Triangle" (or "GURT") as was first introduced by the author in the 2021 referred journal article entitled, "Tri–Power Analysis: The Advanced Post Hoc Triostatistical Assurance Model that Consists of Multiple Advanced Triostatistics to further

Verify, Validate, and Make Viable Ideally Replicated Results of Innovative Investigative Inquiry" in the Journal of Creative Education. It is important to note that the Upright Right Triangle Notation is equitable to the following equation which denotes its intrinsic equality to Gauss' original Eureka Theorem: $[\triangledown + \triangledown + \triangledown = \Delta + \Delta + \Delta = num]$; within the confines of the upright right triangle identity "$[n^2 + n] \div 2$" (is a calculation which equates directly to the traditional equation "Triangle Number Theorem" as: $\frac{n(n+1)}{2}$) is now written as the dual identity:

$$\sum_{i=1}^{n=3} \triangledown_{num} \equiv Tri[n]_{\triangledown} \equiv [n^2 + n] \div 2$$

as

$$\sum_{i=1}^{n=3} \triangledown_{num} = Tri[n]_{\triangledown} = n[n + 1] \div 2 = [num] = [n];$$ where, "$n(n + 1)/2$" = "Triangle Number Equation"; and

"$\sum_{i=1}^{n=3} \triangledown_{num} = Tri[n]_{\triangledown} = n[n + 1] \div 2 = [num] = [n]$" is the Upright Right "Triangle Number Theorem".

Table 1 follows and illustrates integers 1 through 5 as "Upright Right Triangle Numbers".

Table 1
The Upright Right Triangle Positive Integer Table Exhibiting Values 1 to 5

The Tabular Proof of the above as an adaptation of Knott's 2003 Triangle Number Table that Presents the Triangle Numbers in the Golden Upright Right Triangle Model Form

[n] = number	1	2	3	4	5	...
Tri[n] as a sum	1	1+2	1+2+3	1+2+3+4	1+2+3+4+5	...
All numbers 2 and higher expressed as Tri[n] as Upright Right Triangles or Tri[n]	•	•• •	••• •• •	•••• ••• •• •	••••• •••• ••• •• •	...
Final Answer to: Tri[n] = [n² + n]÷2	1	3	6	10	15	...

All numbers or integers have a base trichotomy, for example the number 1 has: +1; -1; and non-1 or zero. In addition, all numbers or integers are triangular in their separate trichtomies as illustrated in the Table above (based upon the original work of Knott, 2003).

The Upright Right Triangle Positive Integer Grid Exhibiting Values 1 to 28 [The Upright Right Triangle Number Grid—Adding Positive Integer According to Base Column Number Across Rows that Creates the Following Table]

↓28	↓21	↓15	↓10	↓6	↓3	←+1
↑2?	↑20	↑14	↑9	↑5	←↑+2	—
↑2?	↑18	↑12	↑7	←↑+3	—	—
↑2?	↑15	↑9	←↑+4	—	—	—
↑1?	↑11	←↑+5	—	—	—	—
↑1?	←↑+6	—	—	—	—	—
↑?	—	—	—	—	—	—

Note that the shape of the numbers above fall into the overall form of the open Upright Right Triangle.

All Positive Integers including Zero as Upright Right Triangle Numbers

All positive integers including zero "0" are "Upright Right Triangle Numbers" and their respective values can all be trichotomously created in a trifold manner via a threefold Tripositive calculation. The Tripositive calculation is satisfied by the equation: $[(n^2 + n)/2]$, which is the mathematical formula used to create the Triangular Numbers need to complete all positive integers including zero as "Upright Right Triangle Numbers". The Tripositive is... The Tripositive is written in the following manner: "[value]". In the case of the "Upright Right Triangle Numbers", the Tripositive number is written as "[num]" and is parsimoniously rewritten as, "[n]", thus, "[num] = [n]", for all positive integers including zero ("0").

In deference to the aforementioned, the following is then true for the "Positive Triangular Number Values" that are used to determine all numbers as "Upright Right Triangle Numbers" are indicated as follows using the "Upright Right Triangle Number Equation" of "$[(n^2 + n)/2]$" (using zero "0" and the positive integers 1 through 9 in the "Upright Right Triangle Number Equation") are presented in the Table below.

The Positive Upright Right Triangle Integer Table Exhibiting Values 0 to 28

"Positive Right Triangle Values" (including Zero) for the Triangular Numbers Equations	The "Upright Right Triangle Number Equations" for the "Positive Right Triangle Values"	Final Numerical Outcomes
For, n = 0	Then, $0 = [(n^2 + n)/2] = [(0^2 + 0)/2]$	$0/2 = 0$
For, n = 1	Then, $1 = [(n^2 + n)/2] = [(1^2 + 1)/2]$	$2/2 = 1$
For, n = 3	Then, $3 = [(n^2 + n)/2] = [(2^2 + 2)/2]$	$6/2 = 3$
For, n = 6	Then, $6 = [(n^2 + n)/2] = [(3^2 + 3)/2]$	$12/2 = 6$
For, n = 10	Then, $10 = [(n^2 + n)/2] = [(4^2 + 4)/2]$	$20/2 = 10$
For, n = 15	Then, $15 = [(n^2 + n)/2] = [(5^2 + 5)/2]$	$30/2 = 15$
For, n = 21	Then, $21 = [(n^2 + n)/2] = [(6^2 + 6)/2]$	$42/2 = 21$
For, n = 28	Then, $28 = [(n^2 + n)/2] = [(7^2 + 7)/2]$	$56/2 = 28$
For, n = 36	Then, $36 = [(n^2 + n)/2] = [(8^2 + 8)/2]$	$72/2 = 36$
For, n = 45	Then, $45 = [(n^2 + n)/2] = [(9^2 + 9)/2]$	$90/2 = 45$
	...and the pattern continues...	

The abovementioned Triangular Numbers as the "Final Numerical Outcomes" in the Table can then be used to repetitively create all positive numbers in groups of 3 to illustrate that all positive numbers (positive integers) are Triangular Numbers. Table 4 follows and illustrates the

The Upright Right Triangle Positive Integer Table Exhibiting Values 0 to 45

"Positive Integer" (including Zero)	The "Positive Right Triangle Value Equations" Using the "Positive Right Triangle Values"	Final Positive Integer
For, n = 0	[0] + [0] + [0] = 0	0
For, n = 1	[0] + [0] + [1] = 1	1
For, n = 2	[0] + [1] + [1] = 2	2
For, n = 3	[1] + [1] + [1] = 3	3
For, n = 4	[0] + [1] + [3] = 4	4
For, n = 5	[1] + [1] + [3] = 5	5
For, n = 6	[0] + [3] + [3] = 6	6
For, n = 7	[1] + [3] + [3] = 7	7
For, n = 8	[1] + [1] + [6] = 8	8
For, n = 9	[0] + [3] + [6] = 9	9
For, n = 10	[0] + [0] + [10] = 10	10
For, n = 11	[0] + [1] + [10] = 11	11
For, n = 12	[1] + [1] + [10] = 12	12
For, n = 13	[0] + [3] + [10] = 13	13
For, n = 14	[1] + [3] + [10] = 14	14
For, n = 15	[0] + [0] + [15] = 15	15
For, n = 16	[0] + [1] + [15] = 16	16
For, n = 17	[1] + [1] + [15] = 17	17
For, n = 18	[0] + [3] + [15] = 18	18
For, n = 19	[1] + [3] + [15] = 19	19
For, n = 20	[0] + [10] + [10] = 20	20
For, n = 21	[0] + [0] + [21] = 21	21
For, n = 22	[0] + [1] + [21] = 22	22
For, n = 23	[1] + [1] + [21] = 23	23
For, n = 24	[0] + [3] + [21] = 24	24
For, n = 25	[1] + [3] + [21] = 25	25
For, n = 26	[1] + [10] + [15] = 26	26
For, n = 27	[6] + [6] + [15] = 27	27
For, n = 28	[0] + [0] + [28] = 28	28
For, n = 29	[0] + [1] + [28] = 29	29
For, n = 30	[0] + [15] + [15] = 30	30
For, n = 31	[0] + [3] + [28] = 31	31
For, n = 32	[1] + [3] + [28] = 32	32
For, n = 33	[6] + [6] + [21] = 33	33
For, n = 34	[0] + [6] + [28] = 34	34
For, n = 35	[1] + [6] + [28] = 35	35
For, n = 36	[0] + [0] + [36] = 36	36
For, n = 37	[0] + [1] + [36] = 37	37
For, n = 38	[1] + [1] + [36] = 38	38
For, n = 39	[0] + [3] + [36] = 39	39
For, n = 40	[6] + [6] + [28] = 40	40
For, n = 41	[3] + [10] + [28] = 41	41
For, n = 42	[0] + [6] + [36] = 42	42
For, n = 43	[0] + [15] + [28] = 43	43
For, n = 44	[6] + [10] + [28] = 44	44
For, n = 45	[0] + [0] + [45] = 45	45
	...and the pattern continues...	

In summary and deference to the aforementioned, the following is then true for the "Positive Triangular Number Values" that are used to determine all numbers as "Trichotomous Upright Right Triangle Numbers" are indicated as follows using the "Trichotomous Upright Right Triangle Number Equation" of "$[(n^2 + n)/2]$" (using zero "0" and the positive integers 1 through 9 in the "Trichotomous Upright Right Triangle Number Equation") written in a "non-tabular formulaic format" as follows:

For n = 0, then 0 = $[(n^2 + n)/2]$ = $[(0^2 + 0)/2]$ = 0/2 = 0;
For n = 1, then 1 = $[(n^2 + n)/2]$ = $[(1^2 + 1)/2]$ = 2/2 = 1;
For n = 3, then 3 = $[(n^2 + n)/2]$ = $[(2^2 + 2)/2]$ = 6/2 = 3;
For n = 6, then 6 = $[(n^2 + n)/2]$ = $[(3^2 + 3)/2]$ = 12/2 = 6;
For n = 10, then 10 = $[(n^2 + n)/2]$ = $[(4^2 + 4)/2]$ = 20/2 = 10;
For n = 15, then 15 = $[(n^2 + n)/2]$ = $[(5^2 + 5)/2]$ = 30/2 = 15;
For n = 21, then 21 = $[(n^2 + n)/2]$ = $[(6^2 + 6)/2]$ = 42/2 = 21;
For n = 28, then 28 = $[(n^2 + n)/2]$ = $[(7^2 + 7)/2]$ = 56/2 = 28;
For n = 36, then 36 = $[(n^2 + n)/2]$ = $[(8^2 + 8)/2]$ = 72/2 = 36;
For n = 45, then 45 = $[(n^2 + n)/2]$ = $[(9^2 + 9)/2]$ = 90/2 = 45;

The next series of Triangular Numbers is 10, 15, and 21 respectively to repeat the pattern and continue to define all positive integers as Triangular Numbers. Written in a "non-tabular full format" as follows:

Where, n = 10, then 10 = $[(n^2 + n)/2]$ = $[(4^2 + 4)/2]$ = 20/2 = 10; and n = 15, then 15 = $[(n^2 + n)/2]$ = $[(5^2 + 5)/2]$ = 30/2 = 15; onto n = 21, then 21 = $[(n^2 + n)/2]$ = $[(6^2 + 6)/2]$ = 42/2 = 21; and the pattern continues...

In summary, the Positive Triangular Numbers are calculated as:

n = 0, then 0 = [(n² + n)/2] = [(0^2 + 0)/2] = 0/2 = 0;
n = 1, then 1 = [(n² + n)/2] = [(1^2 + 1)/2] = 2/2 = 1;
n = 3, then 3 = [(n² + n)/2] = [(2^2 + 2)/2] = 6/2 = 3;
n = 6, then 6 = [(n² + n)/2] = [(3^2 + 3)/2] = 12/2 = 6;
n = 10, then 10 = [(n² + n)/2] = [(4^2 + 4)/2] = 20/2 = 10;
n = 15, then 15 = [(n² + n)/2] = [(5^2 + 5)/2] = 30/2 = 15;
n = 21, then 21 = [(n² + n)/2] = [(6^2 + 6)/2] = 42/2 = 21; and the pattern continues…

The abovementioned Triangular Numbers can then be used to repetitively create all positive numbers in groups of 3 to illustrate that all positive numbers (positive integers) are Triangular Numbers.

For example:

n = 0, because, [0] + [0] + [0] = 0;
n = 1, because, [0] + [0] + [1] = 1;
n = 2, because, [0] + [1] + [1] = 2;
n = 3, because, [1] + [1] + [1] = 3;
n = 4, because, [0] + [1] + [3] = 4;
n = 5, because, [1] + [1] + [3] = 5;
n = 6, because, [0] + [3] + [3] = 6;
n = 7, because, [1] + [3] + [3] = 7;
n = 8, because, [1] + [1] + [6] = 8;
n = 9, because, [0] + [3] + [6] = 9;
n = 10, because, [0] + [0] + [10] = 10;
n = 11, because, [0] + [1] + [10] = 11;
n = 12, because, [1] + [1] + [10] = 12;
n = 13, because, [0] + [3] + [10] = 13;
n = 14, because, [1] + [3] + [10] = 14;
n = 15, because, [0] + [0] + [15] = 15;

n = 16, because, [0] + [1] + [15] = 16;
n = 17, because, [1] + [1] + [15] = 17;
n = 18, because, [0] + [3] + [15] = 18;
n = 19, because, [1] + [3] + [15] = 19;
n = 20, because, [0] + [10] + [10] = 20;
n = 21, because, [0] + [0] + [21] = 21;
n = 22, because, [0] + [1] + [21] = 22;
n = 23, because, [1] + [1] + [21] = 23;
n = 24, because, [0] + [3] + [21] = 24;
n = 25, because, [1] + [3] + [21] = 25;
n = 26, because, [1] + [10] + [15] = 26;
n = 27, because, [6] + [6] + [15] = 27;
n = 28, because, [0] + [0] + [28] = 28;
n = 29, because, [0] + [1] + [28] = 29;
n = 30, because, [0] + [15] + [15] = 30;
n = 31, because, [0] + [3] + [28] = 31;
n = 32, because, [1] + [3] + [28] = 32;
n = 33, because, [6] + [6] + [21] = 33;
n = 34, because, [0] + [6] + [28] = 34;
n = 35, because, [1] + [6] + [28] = 35;
n = 36, because, [0] + [0] + [36] = 36;
n = 37, because, [0] + [1] + [36] = 37;
n = 38, because, [1] + [1] + [36] = 38;
n = 39, because, [0] + [3] + [36] = 39;
n = 40, because, [6] + [6] + [28] = 40;
n = 41, because, [3] + [10] + [28] = 41;
n = 42, because, [0] + [6] + [36] = 42;
n = 43, because, [0] + [15] + [28] = 43;
n = 44, because, [6] + [10] + [28] = 44;
n = 45, because, [0] + [0] + [45] = 45; and the pattern continues...

A Summative Graph of Tripositive Trichotomous Upright Right Triangle Integers

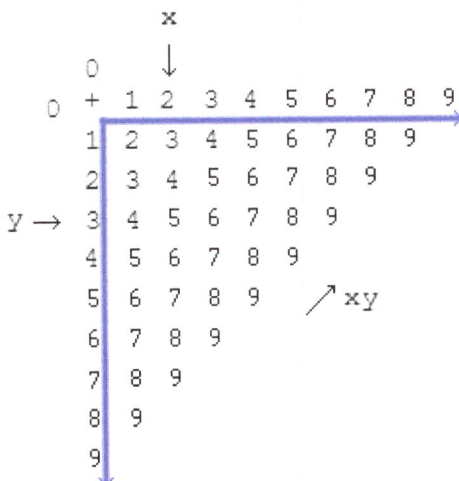

```
              x
      0       ↓
  0   +  1  2  3  4  5  6  7  8  9
      1  2  3  4  5  6  7  8  9
      2  3  4  5  6  7  8  9
y →   3  4  5  6  7  8  9
      4  5  6  7  8  9
      5  6  7  8  9        ↗ xy
      6  7  8  9
      7  8  9
      8  9
      9
```

The Triological Science "Trifold Trichotomous Mathematical Operations"

Triangle Numbers have "Trifold Trichotomous Mathematical Operations" that parallel the Triological Science of Triology that observes nature's subatomic structure of the atom: Positive [as the Proton], Negative [as the Neutron], and Neutral [as the Electron]. The "Trifold Trichotomous Mathematical Operations" have a trichotomy of 3 "▽ Trichotomous Mathematical Operations (▽ Multiplication, ▽ Addition, and ▽ Subtraction) that in summation together create the final "▽Trichotomous Summative Harmonic Equation" that fields the 3-4-5-6 Golden Upright Right Triangle that has multiple definitive trichotomous mathematical qualities, characteristics, and elements. All of the aforementioned is illustrated, calculated, and represented graphically in the illustrated graphs and equations that follow.

Trioengineering Meronymics

Definition: The Meronymic (meaning "parts") Mathematics Operations of the "∇ Trichotomous Mathematical Operation" of the "F Trichotomous Summative Harmonic Equation and Operation" and the Holonymic (meaning "whole") Final "∇ Trichotomous Summative Harmonic Equation and Operation".

The "∇ Trichotomous Meronymic Multiplication Mathematical Equation and Operation" and its Geometric Illustration:

$$\nabla[x \cdot y] = \quad \begin{array}{c|ccccc} \times & 1 & 2 & 3 & 4 & x \\ \hline 1 & 1 & 2 & 3 \nearrow \\ 2 & 2 & 4 \\ 3 & 3 \\ y \end{array}$$

The "∇Trichotomous Meronymic Summation Mathematical Equation and Operation" and its Geometric Illustration:

$$\nabla[x + y] = \quad \begin{array}{c|ccccc} + & 1 & 2 & 3 & 4 & x \\ \hline 1 & 2 & 3 & 4 \nearrow \\ 2 & 3 & 4 \\ 3 & 4 \\ y \end{array}$$

The "∇Trichotomous Meronymic Subtraction Mathematical Equation and Operation" and its Geometric Illustration:

$$\nabla[x-y] = \begin{array}{c|cccc} 0 & - & 1 & 2 & 3 & 4 & x \\ \hline 1 & 0 & 1 & 2 \nearrow \\ 2 & -1 & 0 \\ 3 & -2 \\ \hline & y \end{array}$$

The Final "∇ Trichotomous Holonymic Summative Harmonic Operation" and its Geometric Illustration:

$$\nabla\Big[[x \cdot y] + [x + y] + [x - y]\Big] = \begin{array}{c|cccc} 0 & [\,] & 1 & 2 & 3 & 4 & x \\ \hline 1 & 3 & 6 & 9 \nearrow \\ 2 & 4 & 8 \\ 3 & 5 \\ \hline & y \end{array}$$

All of the ∇ Trichotomous Mathematical Operations in a Geometric Illustration exhibiting the holonymic sequential nature of the Equations that explain, define, lead to, and are a part of the 3-4-5-6 Golden Upright Right Triangle:

$$[x \cdot y] = \qquad\qquad [x + y] = \qquad\qquad [x - y] =$$

0	×	1	2	3	4	x
1	1	2	3↗			
2	2	4				
3	3					
y						

0	+	1	2	3	4	x
1	2	3	4↗			
2	3	4				
3	4					
y						

0	–	1	2	3	4	x
1	0	1	2↗			
2	-1	0				
3	-2					
y						

$$[x \cdot y] + [x + y] + [x - y] =$$

0	[ ]	1	2	3	4	x
1	3	6	9↗			
2	4	8·				
3	5·					
y						

=

=

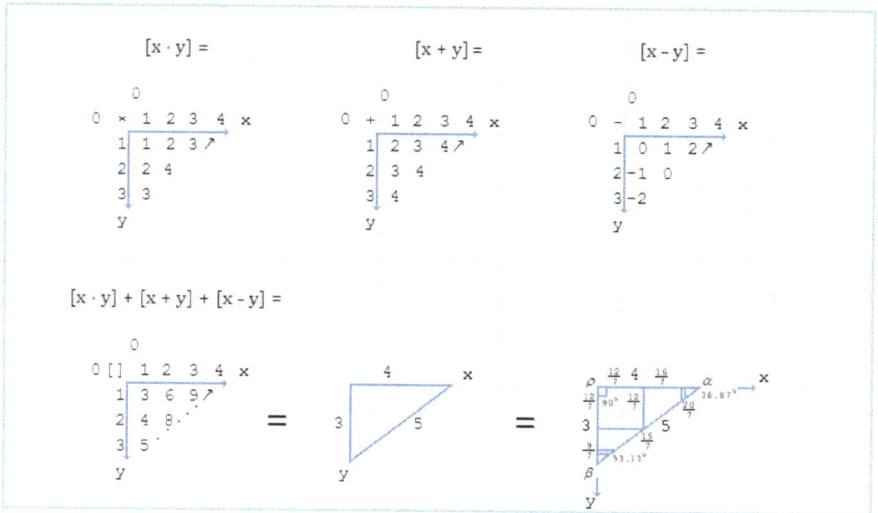

Why the 3-4-5-6 Upright Right Triangle is "Golden"

The uniqueness of the equality of the 3 by 3 Table with the 4 by 3 Table to create the 3-4-5-6 Golden Upright Right Triangle is illustrated in the following graphic and mathematical equations and formulae in the section that follows immediately after expressed using Trioengineering Notation:

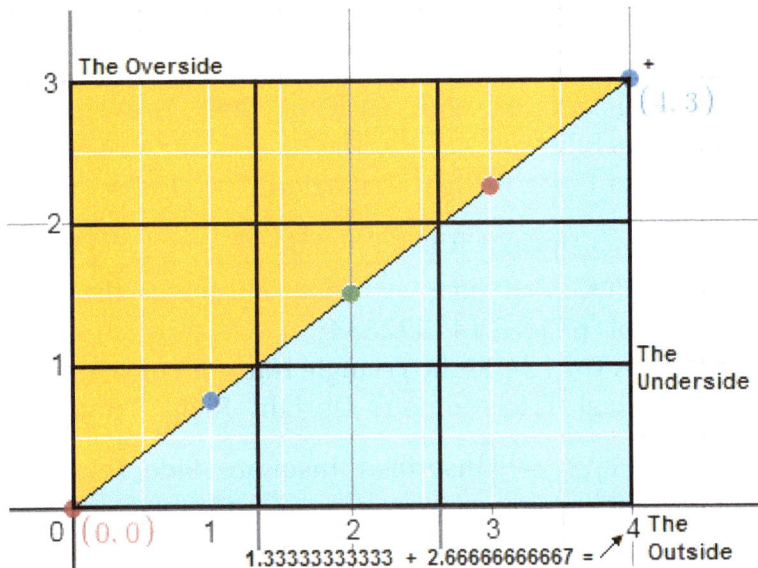

The Overside

(1.3)

The
Underside

0 (0.0)   1   2   3   4 The
1.33333333333 + 2.66666666667 = Outside

The uniquely inherent characteristics as unit measurements of the 3-4-5-6 Trichotomous Upright Right Triangle are what makes it "Golden". The calculative characteristics allow for the 3-4-5-6 Trichotomous Upright Right Triangle to create a triune trichotomous unity between three of the special and remarkable mathematical constants that can be found in and derived from nature. They are the "Trichotomous Upright Right Triangle Pi" = " $\nabla\pi$ " = " $\pi_{[\nabla]}$ " (which is the difference the circumference of a circle and its diameter); the "Trichotomous Upright Right Triangle Natural Logarithm" = " ∇e " = " $e_{[\nabla]}$ " (which is the); and

the "Trichotomous Up Right Triangle Golden Ratio" = " $\nabla\varphi$ " = " $\varphi_{[\nabla]}$ " (which is the final value of a given line segment split into two pieces of different lengths so that the longer segment is equal to double the shorter segment). All three of the aforementioned are equal within the confines of the Trichotomous Upright Right Triangle. The "Golden" part of the name of the 3-4-5-6 Upright Right Triangle comes from the "Golden" in the "Golden Triangle Ratio" that is exactly the same as original "Golden Ratio" and is a part of the entire 3-4-5-6 Golden Upright Right Triangle. The precise calculations for each of the three trichotomous calculations with each of their respective original calculations are as follows:

1.) "Trichotomous Upright Right Triangle Pi" = " $\nabla\pi$ " = " $\pi_{[\nabla]}$ " = "The Ideal Trichotomous Upright Right Triangle Pi" = $\pi_{[\blacktriangleright]}$ = $\left[\bar{x}_{[\blacktriangleright]} \div \sqrt{\phi}\right]$ = 3.144605511029693... (which is very close to the traditional numerical value of "pi") $\cong 3.141592654$;

2.) "Trichotomous Upright Right Triangle Natural Logarithm" = " ∇e " = " $e_{[\nabla]}$ " = "The Ideal Trichotomous Upright Right Triangle Natural Logarithm e " = $e_{[\blacktriangleright]}$ = $\left[\text{Inscribed Insquare Side } (s) + \nabla abc\right]$ = 2.714285714285714285... (which is very close to the traditional numerical value of "e") $\cong 2.718281828459045...$; and

3.) "Up Right Triangle Golden Ratio" = " $\nabla\phi$ " = " $\phi_{[\nabla]}$ " = "The Ideal Trichotomous Upright Right Triangle Phi" = $\phi_{[\blacktriangleright]}$ =

$$\left[1 + \sqrt{b\left[\sqrt{1 + \left[\tfrac{a}{b}\right]^2}\right]}\right] \div 2 = 1.618033988749895...\text{ (which is exactly}$$

identical to the traditional numerical value of "phi" = " ϕ ") $\equiv$ 1.618033988749895...

Graphically all of the aforementioned mathematical equations and their equality as identities associated in and as elemental foundational characteristics of the "3-4-5-6 Golden Upright Triangle" can be represented in the following definitive illustration:

x-axis or x-coordinate (abscissa)

$\angle \rho$

$a \equiv 4$

$\angle \alpha$

$90°$ $36.87°$

$b \equiv 3$ $\frac{12}{7}$ $\frac{12}{7}$

$c \equiv 5$

53.13

$\angle \beta$

The Inclination of $c \equiv \left[\frac{3}{4}\right] \equiv 0.75 \equiv ["\nearrow"]$

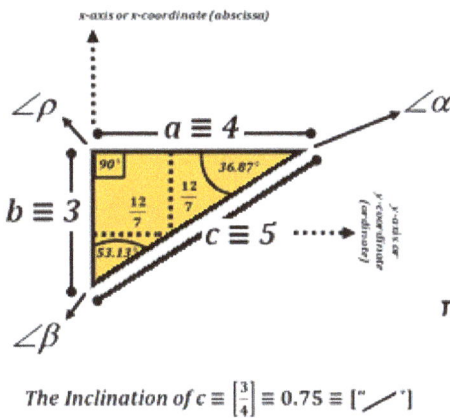

$$\phi_r \equiv \frac{1 + \sqrt{3\sqrt{1 + \left[\frac{4}{3}\right]^2}}}{2}$$

$$c \equiv \left[b\sqrt{1 + \left[\frac{4}{3}\right]^2}\right] \qquad e_r \equiv \left[\frac{12}{7} + 1\right]$$

$$\pi_r \equiv \left[\cfrac{4}{\sqrt{\cfrac{1 + \sqrt{3\sqrt{1 + \left[\frac{4}{3}\right]^2}}}{2}}}\right] \equiv$$

$$\pi_r \equiv \left[\cfrac{a}{\sqrt{\phi_r}}\right] \equiv \pi_r \equiv \left[\cfrac{4}{\sqrt{\phi_r}}\right]$$

The Extracting of $\sqrt{abc}$ from the Visualus Isometric Cuboid

Reducing **xyz** from three coordinates into two coordinates that consist of **x** and **y**:

$$\frac{[xyz]}{[z]} = \frac{[x][y][z]}{[z]} = \frac{[x][y][\cancel{z}]}{[\cancel{z}]} = \frac{[x][y]}{1} = [x][y] = [xy] \text{ as}$$

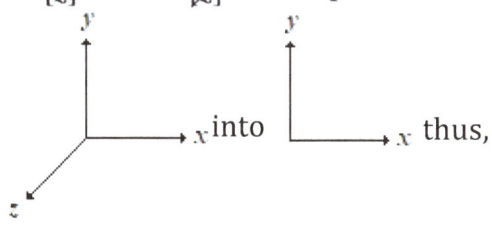

x into   x thus,

$$[abc] = [a][b][c] \text{ into:}$$

$$\frac{[abc]}{[c]} = \frac{[a][b][c]}{[c]} = \frac{[a][b][\cancel{c}]}{[\cancel{c}]} = \frac{[a][b]}{1} = [a][b] = [ab], \text{ therefore } [xy] = [ab],$$

just as **[xyz] = [abc]**.

This then yields the following:

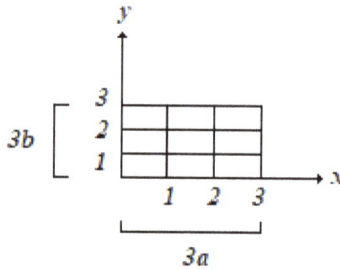

Where, the size (or "magnitude") on the *x*-axis = 4/3 per section (as a distance or measurable length) of each cell. Conversely, the size (or "magnitude") on the *y*-axis = 1 per section (as a distance or measurable height) of each cell. This can be exhibited graphically in the following manner:

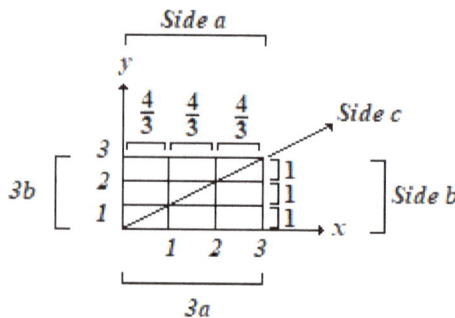

According to the model displayed above: $\frac{4}{3} = 1.333...$, Note: $[1.333... \times 3] = 4$. The 3 by 3 Table Format is composed of 2 Right Triangles one is upright and the other is inverted. This is illustrated in the following manner:

Note: 2 Right Triangles

Note: **[upr]** = ("Upright") indicated by the unshaded Trichotomous Upright Right Triangle and opposite of it is **[inv]** = ("Inverted") indicated by the shaded light-gray area, together the two areas create the Standard 3 by 3 Table that is in tri-coordinates the front face of the Visualus Isometric Cuboid. The measurements of the 3-4-5-6 Golden Upright Right Triangle = " ∇*abc*" are: *Side a* = 4; *Side b* = 3; *Side c* = 5; with a final *Area* of "A" = 6, respectively.

The Geometric Explanation of the " ∇Right Conversion" to create the " ∇Right Triangle Trichotomous Equation Characteristics of Side *c*"

The key to understanding all of the Triological Sciences begins with comprehending their foundation which is the 3-4-5-6 Golden Upright Right Triangle. To ideally grasp all of the meanings and subtle nuances of the Golden Upright Right Triangle one must first understand that it has its origin in the mathematics of Visualus and its foundation which is grounded in the Isometric Cuboid. The Isometric Cuboid is the visual manifestation of the Rectilinear (also known as Linear) Model of Instructional Systems Design. The Front Face of the Isometric Cuboid is where the 3-4-5-6 Golden Upright Right Triangle is derived. The Front Face is a 3 by 3 Table composed of 9 cells that make up the 9 sections of the original Isometric Cuboid. The 3 by 3 Table must be converted or transformed into a 4 by 3 Table of 12 equally proportioned 4 by 3 " 1 by 1 Squares" to begin constructing the three side unit measurements of the Golden Upright Right Triangle. This is referred to as mathematically "Squaring the Cuboid" as it is a mathematical geometric transformation from the Cuboid to a 4 by 3 Table to derive the Golden Upright Right Triangle (alternatively the transformation from the 4 by 3 Table into the full Isometric Cuboid is referred to as "Cubing to the 1 by 1 Square" or more simply called "Cubing the Square"). All of this is possible due to the unique relationship that exists between Sides *a* and *b* respectively of the 3-4-5-6 Golden Upright Right Triangle. The Golden Upright Right Triangle "Side *a*" always has a unit measure of 4 and "Side *b*" always has a unit measure of 3. The slope/incline/acclivity of the Golden Upright Right Triangle when it sections the original 3 by 3 Table into two halves is [∇m] = 1 because, [Side *a* ÷ Side *b*] = 3 ÷ 3 = 1. However, when the "Cubing the Square" transformation into the 4 by 3 Table occurs the slope/incline/acclivity of the Golden Upright Right Triangle now change and becomes [∇m] = 0.75 because, [Side *a* ÷ Side *b*] = 4 ÷ 3 = 0.75. Thus, the same shape can now be represented as two equal ratios to produce the same outcomes in a variety of ways. This allows for the unique features of the original shape and its mathematics to be extended into multiple useful solutions and applications that have great utility in variety of circumstances and makes the 3-4-5-6 Trichotomous Upright Right Triangle a unilaterally useful tool. This is also one of the unique features that makes the 3-4-5-6 Trichotomous Upright Right Triangle "Golden" and initially why it is referred to as the "Golden Upright Right Triangle". The mathematical geometrics of the "Squaring the Cuboid" transformation is illustrated below and occurs in the following manner:

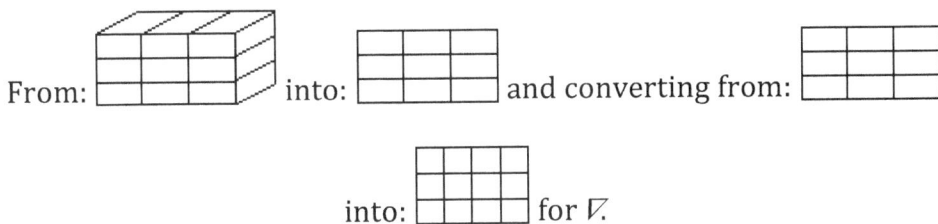

From: into: and converting from:

into: for V.

The mathematics of the above can be presented graphically as:

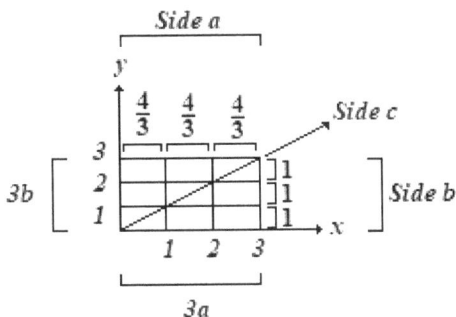

Transforms into the following on a 1 by 1 Square Graph:

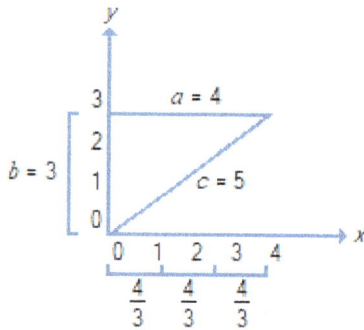

On the graph above: $x = (4, 0)$; and $y =$ Side $b = (0, 3)$.

Explaining how Side $a = 4$ from the "rectangularity" of the 3 by 3 Table (that is the Visualus "Front Face" of the Isometric Cuboid and also the Tri^2 Test 3 by 3 Table Format) can be further illuminated via graphical definition. By defining the equality of the 3 by 3 Table with the 4 by 3 Table by illustrating the equality of the distance of the respective cells, a greater understanding of the conversion and transformation of the 3 by 3 into the 4 by 3 for the unique purposes of measurement can be attained. The illustration below provides the definitive scale measurements of both Tables for comprehension of conversion and transformation in the following manner:

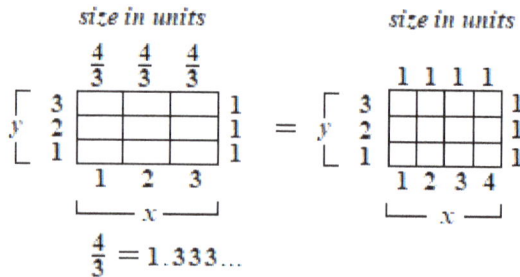

$$\tfrac{4}{3} = 1.333\ldots$$

In the above "Table Equity Conversion and Transformation Illustration" equity is illustrated by the rectangular "3 by 3 Front Face" as one unit on the *y*-axis that has 3 units of 1, while the *x*-axis is equal to 4/3rds = 1.333... into 4 units of 1 on the final 4 by 3 Table. Thus, for the purposes of conversion and transformation that will yield the Pythagorean theorem, extensive general area field dynamics that have greater volumetrics, and Triophysics definitive areas that have broad utility in learning, Triostatistical analytics, and measurable self-growth. Therefore, the 3 by 3 Table is converted and transformed into the 4 by 3 Table in terms of singular cell rectangle units for the purposes of equality from a rectangle into a rectangle that is a 1 by 1 square. This is done by each of the individual cells (as individual singular units on the entire 3 by 3 Table) being sub-divided into Table cell units of a 4 by 3 structure with a total of twelve "1 by 1 Squares" thereby increasing the overall number of cells from 9 into 12 with side lengths changed from 1333... into units of 1. Mathematically, the ratios for the cells of the 4 by 3 Table with the 3 by 3 Table is a difference of 1/3rd that is equal to 1, written as follows:

$$\frac{x}{y} = \frac{4}{3} - \frac{1}{3} = \frac{3}{3} = 1.$$

The Conversion and Transformation from the 3 by 3 Table into the 4 by 3 Table

Graphically the conversion and transformation of the 3 by 3 into the 4 by 3 is illustrated as follows:

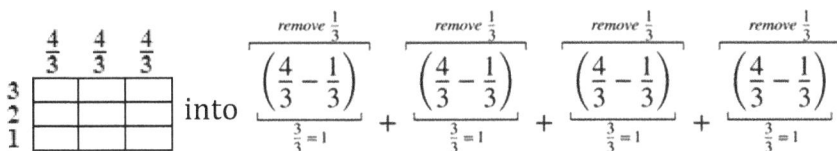

Which then via mathematical geometric conversion and transformation creates the following:

$y = 3$ $\begin{array}{c}3\\2\\1\end{array}$ [3 by 3 grid] $=$ $\begin{array}{c}3\\2\\1\end{array}$ [4 by 3 grid] $y = 3$

$\begin{array}{ccc}1 & 2 & 3\end{array}$ $\begin{array}{cccc}1 & 2 & 3 & 4\end{array}$

$x = 3$ $x = 4$

For each $x = 4/3$, For each $x = 1$

Therefore, the following is the explicative mathematical equation for the above geometric model as a mathematical equation.

The "Side a Conversion and Transformation from the 3 by 3 Table into the 4 by 3 Table" has the following equation:

$\sqrt{}$Side $a = [x_{1...3} - 1/3] + [x_4 = 1/3(3)] = [1 + 1 + 1] + [1] = 4.$

Where,

$x_{1...3} = 4/3$ for x_1, x_2, and x_3.

Thus, the $\sqrt{}$ Side a for $\sqrt{}abc$ Equation is the same as the "Side a Conversion and Transformation Equation" and is:

$\sqrt{}$**Side $a = [x_{1...3} - 1/3] + [x_4 = 1/3(3)] = [1 + 1 + 1] + [1] = 4.$**

In deference to the above, line xy = Side $a = 4$; Side $b = 3$ = The distance of Side a from the x-axis. All of this points on line xy satisfy the condition $y = b$ or y = Side b. Thus, if p(x, y) is any point on line xy, then $y = b$. As such, the equation of Side a as "line a" abc is a rectilinear straight line parallel to the x-axis is $y = b$. The equation of the x-axis for $\sqrt{}abc$ = Tri$_{[abc]}$ is clearly y-axis Side $b = 3$ at the point where Side $a = 4$ is created as line xy, where p(x, y) is any point on the line xy that the endpoint is 4. Thus, the Final Triangle Equation of the Parallel for Side a for $\sqrt{}abc$ = Tri$_{[abc]}$ is: y = Side $b = 3$ or $y = b = 3$. The above graph can be explained and defined numerically (from the abovementioned "Table Equity Conversion and Transformation Illustration") in the following manner:

$$\frac{4}{3} \quad \frac{4}{3} \quad \frac{4}{3}$$

$$0 \quad 1 \quad 2 \quad 3$$

$$0 \quad 1 \quad 2 \quad 3 \quad 4$$

$$\frac{3}{3} \quad \frac{3}{3} \quad \frac{3}{3} \quad \frac{3}{3}$$

$$= \quad = \quad = \quad =$$

$$1 \quad 1 \quad 1 \quad 1$$

The above can be graphically defined this is provided:

The 3 by 3 Table = ⬜ transforms into the 4 by 3 Table = ⬜ that yields the following:

That produces the two Triangles = ⬜ into the two oppositional Triangles = ⬜, the final result being the following:

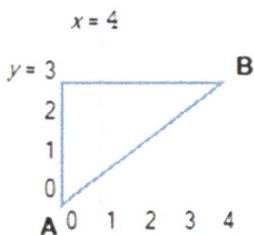

$$x = 4$$

Line **A**(0,0) *to* **B**(4,3)

Thus, 3 by 3 Table Area as the Rectangle Area = A = ∇abc + Δcab =

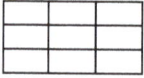

. By removing the inverse of ∇abc as Δcab via subtraction in the following mathematical operation:

∇abc + Δcab

 $\underline{\quad - \Delta cab} =$

∇abc

Where,

$$\nabla abc \;=\; \underset{b}{\overset{a}{\triangle}}\,{}_{c} \quad and \quad \left.\begin{array}{l} a = 4; \\ b = 3; \\ c = 5;\ \text{and} \\ A = 6. \end{array}\right\}$$

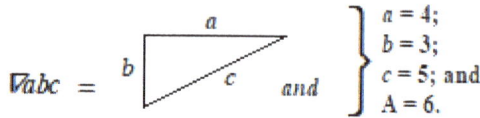

Noting that ∇abc is the "3-4-5-6 Golden Upright Right Triangle" (or "3-4-5-6 GURT"). All of the aforementioned together is the "Foundational Theorem of Triological Science" (also called "The Foundational Theorem of the Triological Sciences"), formulaically written graphically as:

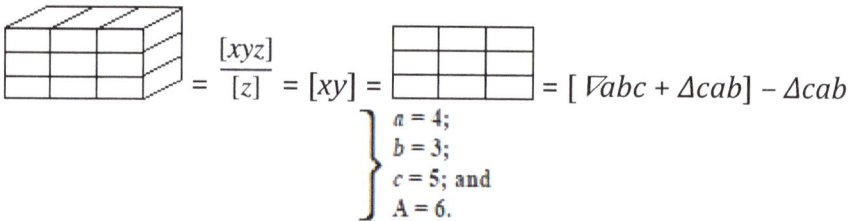

$$= \frac{[xyz]}{[z]} = [xy] = \boxed{\qquad} = [\,\nabla abc + \Delta cab\,] - \Delta cab \;\left.\begin{array}{l} a = 4; \\ b = 3; \\ c = 5;\ \text{and} \\ A = 6. \end{array}\right\}$$

Expressed without geometrics (with full Trines represented) as:

$$\frac{[xyz]}{[z]} = [xy] = [\ \sqrt{abc} + \Delta cab] - \Delta cab = \sqrt{abc} \left.\begin{array}{l} a = 4; \\ b = 3; \\ c = 5; \text{ and} \\ A = 6. \end{array}\right\}$$

Rewritten numerically as:

$$[9xy] \equiv [9ab] \div \tfrac{1}{2}_{[inv]} \equiv \sqrt{abc} \left.\begin{array}{l} a = 4; \\ b = 3; \\ c = 5; \text{ and} \\ A = 6. \end{array}\right\}$$

This is defined in a mathematical geometric identity (that uses the mathematic "identical to" symbol of "$\equiv$" as follows:

$$[9xy] \equiv [9ab] \div \Delta \equiv \sqrt{abc}.$$

Where the above elements of the identity have the following definitions:

$[9xy]$ = The precise Cartesian Coordinates measurements of

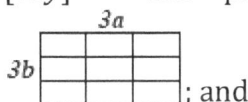

; and

$$[9ab] \div \tfrac{1}{2}_{[inv]} = \quad \text{} \quad - \Delta cab.$$

Where the above elements of the equations presented immediately above have the following further definitions:

$\tfrac{1}{2}_{[inv]} = \Delta cab;$

$\sqrt{abc} =$ that has an Area of "A", where A = 6; and

$\Delta = \underset{a}{\overset{c \quad b}{\triangle}} = \frac{1}{2}_{[inv]} = 0.5_{[inv]}$ (where, "[*inv*]" = inverted which equates to the inverse (or the "upside down and reversed opposite") of the initially presented Trichotomous Upright Right Triangle, that when placed together with original Trichotomous Upright Right Triangle create the original 3 by 3 Table).

This can be rewritten with mathematical parsimony as "The $\sqrt{abc}$ Foundational Theorem of Triological Science (also called "The Triological Science Foundational Theorem of $\sqrt{abc}$") as:

$$\frac{[9ab]}{0.5_{[inv]}} = \sqrt{abc} \equiv \overset{a}{\underset{}{\triangle}}_{b}^{c} \left.\begin{array}{l} a = 4; \\ b = 3; \\ c = 5; \text{ and} \\ A = 6. \end{array}\right.$$

In this manner all of the characteristics and traits of the 3-4-5-6 Golden Upright Right Triangle are inherited by the Trine abc as the representative mathematical symbol of the larger more detailed GURT model as a prime example of mathematical parsimony. Formulaically this can be written via a mathematical equation in the following manner:

$$\sqrt{abc} = \triangle = \overset{a}{\underset{b}{\triangle}}_{c} = \frac{1}{2}_{[upr]} = 0.5_{[upr]}$$ (where, "[*upr*]" = the Trichotomous Upright Right Triangle).

The "∇ Right Triangle y–Intercept Equation" of the "∇ Right Triangle = ∇abc"

The "∇ Right Triangle Inclination of Side c" as the "∇ Right Triangle Acclivity of Side c" is represented by the following equation: [$\nabla m \cdot \nabla x$] = [∇mx], $b = 0$, and [$a/b = \frac{3}{4}$]. This then allows for the "∇ Right Triangle y–Intercept Equation" to be written as: "$\nabla y = [\nabla mx] + b$". To create the GURT the first step requires the Front Face of the Visualus Isometric Cuboid to be transformed through the three Triological Science "Trichotomous Trioengineering Transformations" that are:

(1.) "∇ Right Triangle Trioengineering Intercalation";
(2.) "∇ Right Triangle Trioengineering Inculcation"; and
(3.) "∇ Right Triangle Trioengineering Interpolation".

These aforementioned three are used to create the "Conversion of the Standard 3 by 3 Table" into a 4 by 3 Table for the creation of the Golden Upright Right Triangle with measurements that are: Side $a = 4$, Side $b = 3$, Side $c = 5$, and Area A = 6. Mathematically, the "∇ Right Triangle y–Intercept Equation" is calculated in the following manner: $\nabla y = [\frac{3}{4} \cdot 4] + 0$, therefore, $\nabla y = 3$, thusly, $\nabla m =$ "∇ Right Triangle Slope" = "∇ Right Triangle Inclination" = "∇ Right Triangle Acclivity" = 0.75 the full definition of ∇m is: $\nabla m =$ "∇ Right Triangle Slope" = "∇ Right Triangle Inclination" = "∇ Right Triangle Acclivity" = $\frac{y_2 - y_1}{x_2 - x_1} = \frac{3-0}{4-0} = \frac{3}{4} = 0.75$. The following graphical illustration is true according to the "∇ Right Triangular Interpolation Equation of ∇abc" for the inclination and acclivity of Side c:

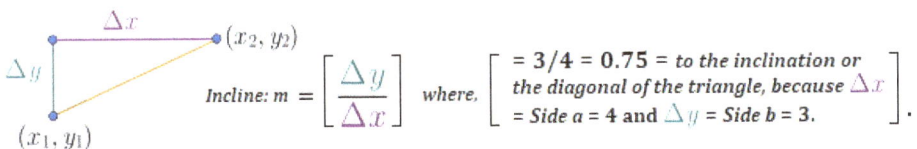

$$\text{Incline: } m = \left[\frac{\Delta y}{\Delta x}\right] \quad where, \quad \left[\begin{array}{l} = 3/4 = 0.75 = to\ the\ inclination\ or \\ the\ diagonal\ of\ the\ triangle,\ because\ \Delta x \\ = Side\ a = 4\ and\ \Delta y = Side\ b = 3. \end{array}\right].$$

Defining and Explaining the Triological Science Trichotomous via the following Three Trichotomous ∇ Right Triological Trioengineering that are:

> **"The ∇Right Triangle Intercalation of ∇abc";**
> **"The ∇Right Triangle Inculcation of ∇abc"; and**
> **"The ∇Right Triangle Interpolation of ∇abc".**

The Triological Science Trioengineering of ∇abc occurs via the trichotomous triune calculative characteristics has three identified distinctive areas that most accurately define the overall uniqueness, utility, and usefulness of the 3-4-5-6 Golden Upright Right Triangle. The three calculative characteristics are viable in that they have interior, exterior, and ulterior measurements that are defined as: (1.) " ∇Right Triangle Intercalation" as the outside cyclical model of ∇abc; (2.) " ∇ Right Triangle Inculcation" as the inside holistic ideation and conceptualization within ∇abc; and (3.) " ∇Right Triangle Interpolation as the .

The Geometric Explanation of the " ∇Right Triangle Intercalation Equation of ∇abc"

Illustrating the "external" (or "exterior") characteristics of the 3-4-5-6 GURT according to the " ∇Triangle Intercalation Equation of ∇abc". The " ∇ Triangle Intercalation Equation" involves the outside of the Triangular Equation Modeling [TEM] as first presented by the author in a 2017 published research paper entitled, "Triangular Equation Modeling [TEM]: The Base Operation, Basic Rationale, and Foundational Logic that is the Basis for "Research Architectural Metrics that are Essential to the Planning, Scope, and Schema of Trichotomous Research Designs" in the i-manager's July–September Journal on Mathematics.

The base [TEM] can be illustrated as a whole as follows (Osler, 2017 in the published research paper entitled, "Triangular Equation Modeling [TEM]: The Base Operation, Basic Rationale, and Foundational Logic that is the Basis for "Research Architectural Metrics" that are Essential to the Planning, Scope, and Schema of Trichotomous Research Designs" that originally appeared in the i-manager's July–September Journal on Mathematics):

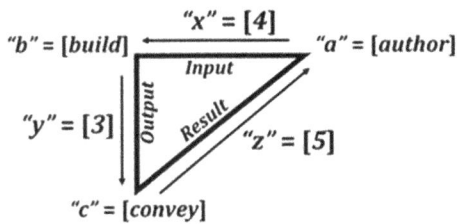

The [TEM] as the Algorithmic Triangular Model (as presented by the author in the published 2021 research paper entitled, "Algorithmic Triangulation Metrics for Innovative Data Transformation:
Defining the Application Process of the Tri–Squared Test" as it in appeared in the April–June i-manager's Journal on Mathematics) is conveyed in the following illustration that illustrates "a trichotomy within a trichotomy" (Osler, 2021) as follows:

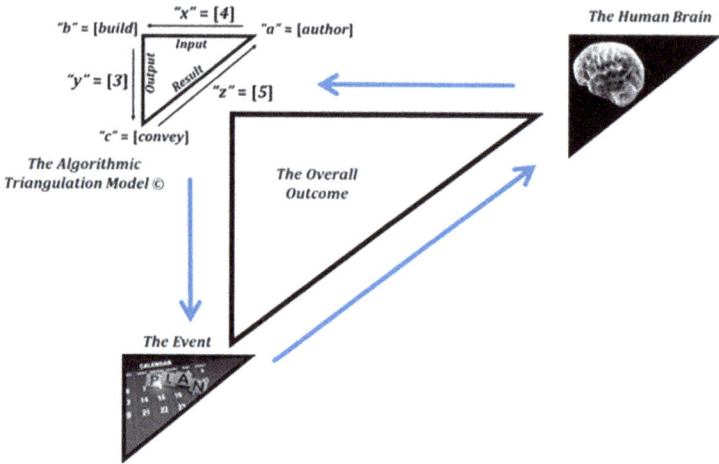

The Algorithmic Triangulation Model ©

The Event

The Overall Outcome

The Human Brain

The [TEM] utilizes the 3-4-5-6 GURT as a trichotomously explicative model to illustrate the unique relationships that exists between, within, and throughout a plethora of algorithms, axioms, concepts, details, ideas, identifications, innovations, inventions, investigations, measurements, principles, solutions, and theories. When used to display the exterior as the outside of the [TEM] then for example the following approach is used:

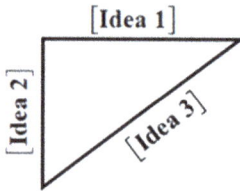

The [TEM] can be further defined contextually in terms of mathematical definitions in the following series of graphics and equations:

The Trichotomous Upright Right Triangle Testing Model Using [TEM] to more accurately describe the Tri–Squared Test and the origins of the Trine Symbol as Originally Published in 2012, 2017, and 2021

The Algorithmic Model of Triangulation (Osler, 2013 as it originally appeared in the referred publication by the author entitled, "Algorithmic Triangulation Metrics for Innovative Data Transformation: Defining the Application Process of the Tri–Squared Test" in the April–June Journal on Mathematics) is of the form:

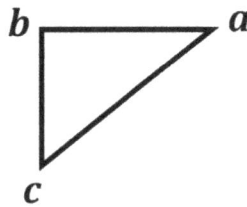

Where,

Vertex **a** = $\nabla\angle a$ = form "Vector $\nabla[a{\to}b]$" as " $\nabla[\overrightarrow{ab}]$" = "**authoring**" = a ratio of a measurement metric of 4 (used to convey information) = The Initial Tri–Squared Instrument Design;

Vertex **b** = $\nabla\llcorner b$ = = form "Vector $\nabla[b{\to}c]$" as " $\nabla[\overrightarrow{bc}]$" = "**building**" = a ratio of a measurement metric of 3 (as the Triostatistics: Tri–Squared Test) = The Tri–Squared Qualitative Instrument Responses which will be analyzed via Tri–Squared Analysis; and

Vertex **c** = $\nabla\angle c$ = = form "Vector $\nabla[c{\to}a]$" as " $\nabla[\overrightarrow{ca}]$" = "**conveying**" = The Final Tri–Squared Test Outcomes in a Quantitative Report.

Thus, the Triangulation Model is symbolized by a Right Triangle written in graphical illustrative format as the following image: "∇" = the italicized nabla as the Trioengineering Notation Trine "∇" that has appeared earlier in this narrative. This symbol is called the "Trine" (meaning a group of three) is written mathematically as "$\nabla = abc$" = "∇abc" and is simplified into the mathematic geometric expression: ∇abc (meaning "Triangulation Model *abc*" or more simply "Trine *abc*" which is the original definition as first published in 2013). The adapted definitions are presented here from that original publication in deference to the Triological Sciences as follows:

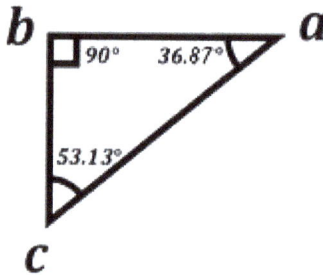

The angles have the following angular measurements in degrees that measure radially as: $\angle a$ = 36.87 degrees, $\angle b$ = 90 degrees, and $\angle c$ = 53.13 degrees respectively, that all add up to the standard 180° of any and all triangles, thus, in terms of the Trioengineering Notation Trine as the GURT is: [36.87° + 90° + 53.13° = 180°]. The connective lines between the Trichotomous Upright Right Triangle vertex points (i.e., the lines between points *a*, *b*, and *c* respectively are geometric "vectors" (lines with both magnitude [or "size"] and direction) making the model a systemic or cyclic process from the point of origin "$\angle a$" back to the original point of origin which is also "$\angle a$". This is illustrated in terms of Cartesian Coordinates as follows:

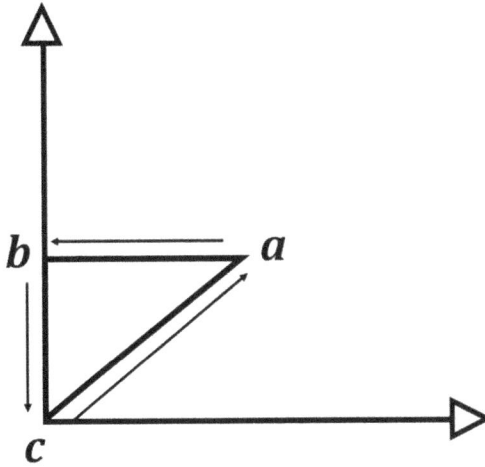

The model above is the beginning of the systemic cyclical methodology of [TEM]. In terms of Vectors, the Trichotomous Upright Right Triangle as the Algorithmic Model of Triangulation now becomes the Trioengineering: Triangulation Model Right Triangle ∇abc is equal to three vectors that illustrate the movement in direction and magnitude from one completed task into another. The entire process is both cyclical and sequential with a "Trine Vector Equation" (using the graphical Trine illustration image) written in Trioengineering Notation as:

$$\nabla = \nabla[\leftarrow x] \rightarrow \nabla[\downarrow y] \rightarrow \nabla[\nearrow z] = \nabla[\nearrow xy], \text{ for } \text{``} \nabla[\overrightarrow{xyz}] \text{''} \text{ as } \text{``} \nabla[\overrightarrow{xyz}] \text{''}$$
$$\text{inbetween } \text{``} \nabla\angle abc\text{''}$$

Defined as Trine = "Concentration of Vector x into Concentration of Vector y into Concentration of Vector z", which is simplified into a more standardized Trine Vector Equation form (using the graphical Trine illustration image) also written in Trioengineering Notation as:

$$\nabla = \nabla\overset{\leftarrow}{x} \rightarrow \nabla\overset{\leftarrow}{y} \rightarrow \nabla\overset{\leftarrow}{z}, \text{ for } \text{``} \nabla\overrightarrow{xyz}\text{''} \text{ as } \text{``} \nabla\overrightarrow{xyz}\text{''} \text{ inbetween } \text{``} \nabla\angle abc\text{''}$$

Where vectors in terms of Trioengineering Notation are: $\nabla \hat{x}$, $\nabla \hat{y}$, and $\nabla \hat{z}$. Where, "z" is not nor has any relation to the Cartesian Coordinate z-axis (also known as the "applicate") here, instead "z" is a recognized vector that is parallel to Side c for the purposes of illustrating the cyclical nature of the 3-4-5-6 Golden Upright Right Triangle as a model (as used in the Triostatistics [TEM] operation and mathematical model). Each of the indicated vectors respectively are indicated on the "Algorithmic Triangulation Data Model" written in a graphical illustration as:

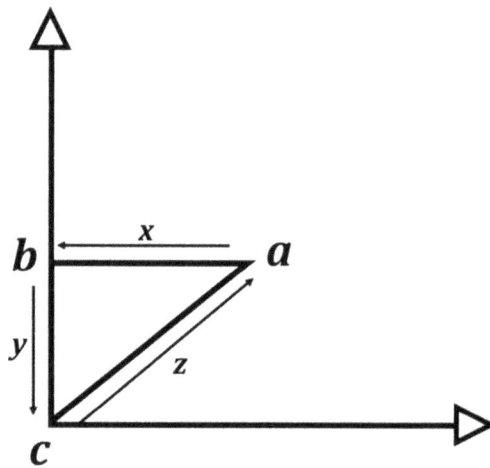

The above illustration indicates that the standardized form of the Trioengineering Notation Trichotomous Vectors in the model above are: $\nabla \hat{x}$, $\nabla \hat{y}$, and $\nabla \hat{z}$ are also equivalent to the Triangle's Sides (a, b, and c) that are the sequential Cartesian Coordinates relative to the size and magnitude of the research engineering phases that sequentially connect the respective angles and report them using Trioengineering Notation as: $\nabla \angle a$, $\nabla \llcorner b$, and $\nabla \angle c$. The aforementioned are written in a sequential cycle illustrating the flow of the vectors in a counterclockwise direction opposite of the Cartesian Coordinates in the first quadrant indicating the more precise direction of arrows from right to left in a motion that can be expressed using Trioengineering Notation as follows:

$$\overleftarrow{\nabla x} = \overleftarrow{\nabla ab};$$
$$\overleftarrow{\nabla y} = \overleftarrow{\nabla bc};\text{ and}$$
$$\overleftarrow{\nabla z} = \overleftarrow{\nabla ac}.$$

Where,

$\nabla z = \nabla xy$ with starting point = (0, 0) to ending point (4, 3) with a slope/incline/acclivity = 0.75.

The Complete Tri–Squared Analysis Algorithmic Triangulation Model as the 3-4-5-6 Trichotomous Upright Right Triangle (adapted and re-edited from the author's publication in 2013 entitled, "Algorithmic Triangulation Metrics for Innovative Data Transformation: Defining the Application Process of the Tri–Squared Test" in the April–June Journal on Mathematics)

The three numeric Vector Operational Phases of the Triangulation Model that now expresses the full scope of the operational parameters of the Triostatistics Tri–Squared Test are completely defined in the following manner (as presented earlier):

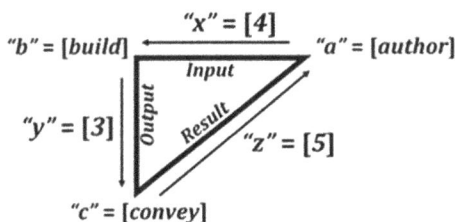

The entire Triangulation Model as a Research Engineering methodology begins with a breakdown of the Trine [∇] Operational

Research Engineering Parameters and Geometric Vectors in the next section.

The Tri–Squared Triangulation Model Research Engineering Process (highlighting the Operational Parameters and Phases of the Tri–Squared Test) for the Triological Sciences

Geometric Vertex ∇a = $\nabla \angle a$ = "**authoring**" = The Initial Tri–Squared Instrument Design = Operational Parameter " ∇a" = "author" = absolute value of ∇a = "modulus ∇a" = $|a|$ = "Trioengineering Notation Trine a" = ∇a = The creation of the Tri–Squared Inventive Investigative Instrument. This process can be seen in the following 3 by 3 cubic model which becomes the 3 by 3 rectangle table model (due to the 4/3rds conversion and transformation mathematics as mentioned in earlier sections of this narrative) of the Standard Tri–Squared Test Data Analysis Table (note: it becomes rectangular due to the top ratio author vector of "4" which represents the four initial sections of the Tri–Squared Test Triple-I instrument as the four Operational Phases of in-depth trichotomous research question data analysis:

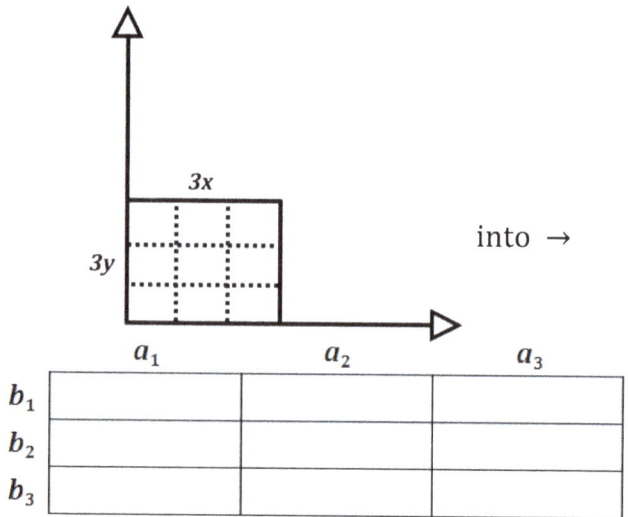

This in turn (written in Trioengineering Notation), leads into [→] vector ∇x = Geometric Vector ∇x = "$\leftarrow \nabla x$" = $\overleftrightarrow{\nabla x}$ = $\nabla \overrightarrow{ab}$ = The Initial Tri–Squared Instrument Construction = Operational Phases "∇x" = absolute value of vector ∇x = "norm ∇x" = $\|x\|$ = "Trine x" = ∇x = The creation of the Tri–Squared Inventive Investigative Instrument = The Pythagorean Triple of the Triangulation Model = "4" = The **4 Phases of Tri–Squared Inventive Investigative Instrument Construction** as "**Trichotomous Psychometrics**" which is composed of 4 Operational Phases that proactively create the "Trioengineered Trichotomous Research Battery" for the Tri–Squared Test. It is this dynamically precise and meticulously detailed trichotomous research battery that is the research investigation "Triple–I" (or "Inventive Investigative Instrument") that is a "Triple Trichotomy" based upon the research investigation's initial three trichotomous research questions. The 4 Phases that attribute to the creation of the "Triple–I" are as follows:

1. a_0 = The Instrument Name (Asset Security, generally in the form of Copyright, or Creative Commons, and/or Trademark);

2. a_1 = Section One of the Research Instrument. Constructed from the first series of instrument items (a. through c.) derived from the research investigation questions as the Qualitative Trichotomous Categorical Variables (as the Initial Investigation Input Variables), evaluated via the Qualitative Trichotomous Outcomes (as the Resulting Outcome Output Variables = b_1, b_2, and b_3 respectively);

3. a_2 = Section Two of the Research Instrument. Constructed from the second series of instrument items (d. through f.) derived from the research investigation questions as the Qualitative Trichotomous Categorical Variables (as the secondary Investigation Input Variables),

evaluated via the Qualitative Trichotomous Outcomes (as the Resulting Outcome Output Variables = b_1, b_2, and b_3 respectively); and

4. a_3 = Section Three of the Research Instrument. Constructed from the third series of instrument items (g. through i.) derived from the research investigation questions as the Qualitative Trichotomous Categorical Variables (as the tertiary Investigation Input Variables), evaluated via the Qualitative Trichotomous Outcomes (as the Resulting Outcome Output Variables = b_1, b_2, and b_3 respectively);

The following table provides the metrics for the construction of the Inventive Investigative Instrument following the parameters indicated in phases 1. through 4. of the first vector [$\overline{x} = \overleftarrow{ab} = 4$] of the Triangulation Model :

$a_0 =$	**"Inventive Investigative Instrument" (or "Triple–I")** —Name— [Adding Asset Security as Copyright = © and/or Trademark = ™]			
$a_1 =$	Section 1. Research Question 1. [Trichotomous Research Battery: One] The First Series of Questions from the Qualitative Trichotomous Categorical Variables are listed below in Items a. through c.			
	Responses: [Select only one from the list.] ▶	b_1	b_2	b_3
	a. Item One based upon Research Question 1	☐	☐	☐
	b. Item Two based upon Research Question 1	☐	☐	☐
	c. Item Three based upon Research Question 1	☐	☐	☐
$a_2 =$	Section 2. Research Question 2. [Trichotomous Research Battery: Two] The Second Series of Questions from the Qualitative Trichotomous Categorical Variables are listed below in Items d. through f.			
	Responses: [Select only one from the list.] ▶	b_1	b_2	b_3
	d. Item Four based upon Research Question 2	☐	☐	☐
	e. Item Five based upon Research Question 2	☐	☐	☐
	f. Item Six based upon Research Question 2	☐	☐	☐
$a_3 =$	Section 3. Research Question 3. [Trichotomous Research Battery: Three] The Third and Final Series of Questions from the Qualitative Trichotomous Categorical Variables are listed below in Items g. through i.			
	Responses: [Select only one from the list.] ▶	b_1	b_2	b_3
	g. Item Seven based upon Research Question 3	☐	☐	☐
	h. Item Eight based upon Research Question 3	☐	☐	☐
	i. Item Nine based upon Research Question 3	☐	☐	☐

The mathematics of the "Trioengineered Tri–Squared Test Triple–I Template" illustration presented above are as follows (in terms of Trioengineering Notation equations) in the next series of equations.

The Complete Trioengineering Tri–Squared Test Triple–I as a "Psychometric Research Trichotomous Battery Template Equation" is:

$$\overset{3}{\underset{i=1}{V}}[a_0 + a_1 + a_2 + a_3].$$

TRIOENGINEERING ™ © *The Problem-Solving Triological Science: The In-Depth Trichotomous Science of the Dynamic 3-4-5-6 Golden Upright Right Triangle for Innovative Problem-Solving.* Osler Studios Incorporated ©, © Copyright 2022 All Rights Reserved.

The 4 Phase Construction Equations for the "Trioengineered Tri–Squared Test Triple–I Template":

$$\nabla a_0 = \nabla[\text{Tri--I}_{[name]} + \text{AS}_{[©; ™]}];$$
$$\nabla a_1 = \nabla[\text{TCV}_{[1]} = \text{TCV}_{1 = \text{Items: } [a....c.]} \rightarrow \text{TOV}_{\text{Tri} = [b_1...b_3]}];$$
$$\nabla a_2 = \nabla[\text{TCV}_{[2]} = \text{TCV}_{2 = \text{Items: } [d....f.]} \rightarrow \text{TOV}_{\text{Tri} = [b_1...b_3]}]; \text{ and}$$
$$\nabla a_3 = \nabla[\text{TCV}_{[3]} = \text{TCV}_{3 = \text{Items: } [g....i.]} \rightarrow \text{TOV}_{\text{Tri} = [b_1...b_3]}].$$

Where, each of the above are mathematically defined below by the individual **"4 Phases of Construction"** in the following manner:

The elements of **Phase One** are defined as follows:
∇a_0 = The Trioengineered Triple–I ("Inventive Investigative Instrument") Name;
$\nabla[\text{Tri--I}_{[name]}];$ = The Trioengineered "Triple–I Name" assigned to this psychometric instrument; and
$\nabla[\text{AS}_{[©; ™]}];$ = The Trioengineered Asset Security with Copyright = © and Trademark = ™.

The elements of **Phase Two** are defined as follows:
∇a_1 = The Trioengineered First Trichotomous Battery based upon the First Research Question as the First Trichotomous Categorical Variable (TCV) and its associated Sub-Questions as Items a., b., and c. respectively with the responses as the Trichotomous Outcome Variables (TOVs) that are b_1, b_2, and b_3 respectively;
$\nabla[\text{TCV}_{[1]}]$ = The Trioengineered Trichotomous Categorical Variable 1;
$\nabla[\text{TCV}_{1 = \text{Items: } [a....c.]}]$ = The Trioengineered (TCV 1) Sub-Questions as Items a., b., and c.;
$\nabla[\rightarrow]$ = The Trioengineered moving forward to Trichotomous Outcome Variables (TOVs); and

$\nabla[\text{TOV}_{\text{Tri} = [b_1 \dots b_3]}]$ = The Trioengineered (TOVs) that are Trichotomous Outcomes b_1, b_2, and b_3.

The elements of **Phase Three** are defined as follows:
∇a_2 = The Trioengineered Second Trichotomous Battery based upon the Second Research Question as the Second Trichotomous Categorical Variable (TCV) and its associated Sub-Questions as Items d., e., and f. respectively with the responses as the Trichotomous Outcome Variables (TOVs) that are b_1, b_2, and b_3 respectively;
$\nabla[\text{TCV}_{[2]}]$ = The Trioengineered Trichotomous Categorical Variable 2;
$\nabla[\text{TCV}_{2 = \text{Items: } [d \dots f.]}]$ = The Trioengineered (TCV 2) Sub-Questions as Items d., e., and f.;
$\nabla[\rightarrow]$ = The Trioengineered moving forward to Trichotomous Outcome Variables (TOVs); and
$\nabla[\text{TOV}_{\text{Tri} = [b_1 \dots b_3]}]$ = The Trioengineered (TOVs) that are Trichotomous Outcomes b_1, b_2, and b_3.

The elements of **Phase Four** are defined as follows:
∇a_3 = The Trioengineered Three Trichotomous Battery based upon the Three Research Question as the Three Trichotomous Categorical Variable (TCV) and its associated Sub-Questions as Items g., h., and i. respectively with the responses as the Trichotomous Outcome Variables (TOVs) that are b_1, b_2, and b_3 respectively;
$\nabla[\text{TCV}_{[3]}]$ = The Trioengineered Trichotomous Categorical Variable 3;
$\nabla[\text{TCV}_{3 = \text{Items: } [g \dots i.]}]$ = The Trioengineered (TCV 3) Sub-Questions as Items g., h., and i.;
$\nabla[\rightarrow]$ = The Trioengineered moving forward to Trichotomous Outcome Variables (TOVs); and

$\nabla[\text{TOV}_{\text{Tri}=[b_1...b_3]}]$ = The Trioengineered (TOVs) that are Trichotomous Outcomes b_1, b_2, and b_3.

The ∇ [TEM] for the "Trioengineered Tri–Squared Test Triple–I Template" illustration is represented geometrically in a Triostatistics [TEM] in the following manner:

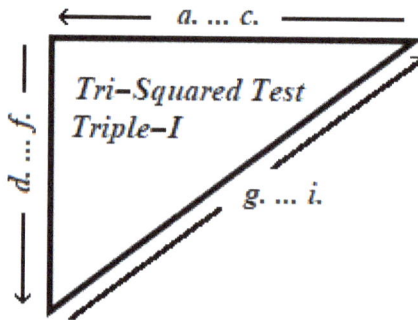

The **Mathematics of the Triangular Equation Modeling [TEM] in Trioengineering Notation Illustrating How the Triostatistics [TEM] comes out of the 3-4-5-6 Golden Upright Right Triangle as the ∇ [TEM] and How the In-Depth Measurements of the Golden Upright Right Triangle Actively and Geometrically Create the 3 by 3 Tri–Square Analysis Table**

The traditional Triostatistics "[TEM]" is equal to the "∇ [TEM]" that emphasizes the connection to the "Trichotomous Upright Right Triangle" via the [TEM] originating from the 3-4-5-6 Golden Upright Right Triangle. The exterior of the [TEM] in deference to its use of the 3-4-5-6 Golden Upright Right Triangle expresses via Trioengineering Notation the Trichotomous Upright Right Triangle Perimeter as "∇P". The ∇P is mathematically expressed in the following series of mathematical formulae and calculations:

$$\nabla P = a + b + \sqrt{a^2 + b^2} = a + b + c$$

$$4 + 3 + \sqrt{4^2 + 3^2} = 7 + \sqrt{16 + 9} = 7 + \sqrt{25}$$

$$7 + 5 = 12, \text{ and}$$

$$[a + b + c] = 4 + 3 + 5 = 12.$$

Therefore, the full numerical Trichotomous Upright Right Triangle Perimeter = The " ∇[TEM] Perimeter" is:

$$\nabla P = 12.$$

The Intercalation of ∇abc in terms of the Trichotomous Upright Right Triangle Perimeter can be rewritten to illustrate the equity of the 3-4-5-6 Golden Upright Right Triangle Sides with the Cartesian Coordinates in the following manner:

$$\nabla P = a + b + c = x + y + z$$

Where,

$\nabla a = a = x$;
$\nabla b = b = y$; and
$\nabla c = c = z$.

Holistically, the "Total Trichotomous Triune Calculative Characteristics of ∇abc" are expressed in a mathematical perimeter definition as follows:

$$\nabla P \text{ of } \nabla abc = \text{Side } a + \text{Side } b + \text{Side } c = 4 + 3 + 5 = 12,$$

Where,

$$\left.\begin{array}{l} \\ \\ \\ \sqrt{abc} \end{array}\right\} \begin{array}{l} a = 4; \\ b = 3; \\ c = 5; \text{ and} \\ A = 6. \end{array}$$

The Geometric Explanation of the "$\sqrt{}$ Right Triangle Inculcation Equation of $\sqrt{abc}$"

Illustrating the "internal" (or "interior") characteristics of the 3-4-5-6 GURT according to the "$\sqrt{}$ Triangle Inculcation Equation of $\sqrt{abc}$". In this manner the $\sqrt{}$ Triangle Inculcation uses the internal holistic characteristic of the [TEM]. Where, $\sqrt{}$ Triangle Intercalation is concerned only with the interior of the [TEM] that illustrates the whole concept, idea, solution and/or problem under observation, investigation and/or study.

Thus, the $\sqrt{}$ Triangle Intercalation as the [TEM] represented as the Algorithmic Triangular Model only displays the interior "Overall Outcome" as the holistic idea, concept, thought and/or problem exhibited as:

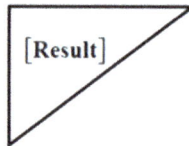

[Result]

The "Result" as entered in the [TEM] above is also equal to the Area of the entire which is A = 6 as illustrated in the [TEM] below.

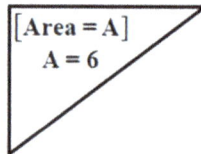

[Area = A]
A = 6

The interior of the [TEM] in deference to its use of the 3-4-5-6 Golden Upright Right Triangle expresses the Trichotomous Upright Right Triangle Area as "∇A". The ∇A is mathematically expressed in the following series of mathematical formulae and calculations:

$$\nabla A = \frac{1}{2}ab = \frac{ab}{2} = \frac{[\text{Side }a][\text{Side }b]}{2} = \frac{4 \cdot 3}{2} = \frac{12}{2} = 6$$

therefore,

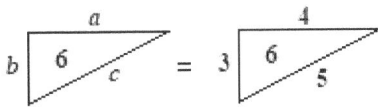
Where, ∇A = A = Area = 6.

Therefore, the full numerical Trichotomous Upright Right Triangle Area = The "∇[TEM] Area" is:

$$\nabla A = 6.$$

The Inculcation of ∇abc in terms of the Trichotomous Upright Right Triangle Area can be rewritten to illustrate the equity of the 3-4-5-6 Golden Upright Right Triangle Sides with the Cartesian Coordinates in the following manner:

$$\nabla A = [ab]/2 = [ab] \div 2 = [xy]/2 = [xy] \div 2$$

because

$$[ab]/2 = [xy]/2 \text{ and } [ab] \div 2 = [xy] \div 2$$

Where,

$\nabla a = a = x$; and
$\nabla b = b = y$.

Holistically, the "Total Trichotomous Triune Calculative Characteristics of ∇abc" are expressed in a mathematical area definition as follows:

$$\nabla A \text{ of } \nabla abc = [\text{Side } a \cdot \text{Side } b] \div 2 = [4 \cdot 3] \div 2 = 6,$$

Where,

$$\nabla abc \left.\begin{array}{l} \\ \\ \\ \\ \end{array}\right\} \begin{array}{l} a = 4; \\ b = 3; \\ c = 5; \text{ and} \\ A = 6. \end{array}$$

The Geometric Explanation of the "∇ Triangle Interpolation Equation for Side c" using Trioengineering Notation

Illustrating the "ulterior" characteristics of the 3-4-5-6 GURT according to the "∇Triangle Interpolation Equation of ∇abc" (also known as the geometric "∇Triangle Interpolation Equation for Side c"):

$$\nabla y = y_1 + (\nabla x - x_1)\frac{(y_2 - y_1)}{(x_2 - x_1)}$$

Where, $(x_1, y_1) = (0, 0)$ for the initial intercept origin point "b", thus, $b = 0$, because, ∇abc rests exactly on the y–axis (the ordinate) and ∇abc is located in the 1st Quadrant (∇positive$_{[x]}$, ∇positive$_{[y]}$) of the Cartesian Coordinates: "⌐I.⌐", as a true shape that has tangible magnitude (size) and distance as it is an exact part of the "Visualus Isometric Cuboid" (that naturally has magnitude and distance as a tri–coordinate shape and form) the "b" therefore is the "y–intercept" as the origin point for $\nabla abc = (x_0, y_0)$, because Side b is on the y–axis, thus, the "∇Triangular Slope–Intercept Equation" is: $\nabla[\nabla y = \nabla m \nabla x + \nabla b] = \nabla[\nabla y = \nabla mx +$

∇b], defined mathematically and geometrically (using Trioengineering Notation) as:

$$\nabla[y = \nabla mx + \nabla b],$$

$$\text{Note: } \nabla[y = \nabla mx + 0],$$

$$\text{Thusly, } \nabla[y = \nabla mx], \text{ because,}$$

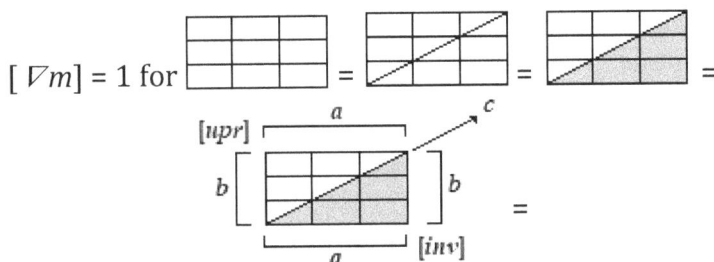

$[\nabla m] = 1$ for

$[upr]$

$[inv]$

$$\frac{\Delta[\nabla y]}{\Delta[\nabla x]} = \frac{\text{Changes in Triangular Coordinate } y}{\text{Changes in Triangular Coordinate } x} = \frac{(y_2 - y_1)}{(x_2 - x_1)} = \frac{(y_2 - 0)}{(x_2 - 0)} = \frac{y_2}{x_2} = \frac{\text{same number}}{\text{same number}} = \frac{\nabla rise}{\nabla run}$$

$$= 1,$$

As such $[m = 1]$ can then be substituted yielding the following: $\nabla[y = (1)x]$, thus, $[\nabla y = \nabla x]$ with a slope of "1" (within the confines of the "3 by 3 Standard Table Format"), therefore representing an exact one to one ratio for $x_{0...3}$ to $y_{0...3}$ or a "1:1 ratio" for $x_{0...3}$ exactly matching $y_{0...3}$ for all points that construct "*Side c*" of "∇abc". Note that "$\nabla b = 0$" as the "∇y–intercept" because Side b rests precisely on the y-axis.

Thusly,

The "Side c ∇Right Trichotomous Triangle Interpolation Equation Interpretations"

The Final "Side c ∇Triangle Interpolation Equation" is written using Trioengineering Notation in the following manner:

$$\text{Side } c = \ \nabla y = 0 + (\nabla x - 0)\frac{(y_2 - 0)}{(x_2 - 0)} \ = \ \nabla y = \nabla x, \text{ as such, } \Delta[\ \nabla y] = \Delta[\ \nabla x] \text{ as a}$$

direct 1:1 ratio for ◿.

Furthermore, there are according to the aforementioned "∇Triangle Interpolation Equation for Side c" trichotomously three separate "interpretations" of the points that construct *Side* c that are respectively expressed as follows:

Interpretation 1.

A grand total of four points for ∇abc according to the "∇Triangle Interpolation Equation Side c". According to the aforementioned Equation, for "b" as the "$y-intercept$" as the origin point for ∇abc = ∇ (x_0, y_0) = when, then, $\nabla y = 0$; and as a result, the points that immediately follow are written using Trioengineering Notation as follows:

$\nabla(x_1, y_1)$ = when, $\nabla x_1 = 1$ then, $\nabla y_1 = 1$;
$\nabla(x_2, y_2)$ = when, $\nabla x_2 = 2$ then, $\nabla y_2 = 2$; and
$\nabla(x_3, y_3)$ = when, $\nabla x_3 = 3$ then, $\nabla y_3 = 3$.

Interpretation 2.

There are only three trichotomous positive points on the positive "incline" and "acclivity" of "Side c" that are represented in the following manner [$\nabla(x_0, y_0)$ is not included as it is neutral and not a positive integer]:

$\nabla(x_1, y_1) =$ when, $\nabla x_1 = 1$ then, $\nabla y_1 = 1$;
$\nabla(x_2, y_2) =$ when, $\nabla x_2 = 2$ then, $\nabla y_2 = 2$; and
$\nabla(x_3, y_3) =$ when, $\nabla x_3 = 3$ then, $\nabla y_3 = 3$.

Interpretation 3.

Lastly, there are two points in the " ∇Triangular Interpolation Equation of Side *c*" (excluding the end points at) that are located in-between the end points of (x_0, y_0) and (x_3, y_3) respectively as a true interpolation of Side *c*:

$\nabla(x_1, y_1) =$ when, $\nabla x_1 = 1$ then, $\nabla y_1 = 1$; and
$\nabla(x_2, y_2) =$ when, $\nabla x_2 = 2$ then, $\nabla y_2 = 2$.

It is important to note the following regarding slope/inclination/acclivity as "[∇m]" respective to the 3 by 3 Table and its conversion and transformation (by "Squaring the Cube") into the 4 by 3 Table geometrically illustrated as follows:

Note: 2 Right Triangles

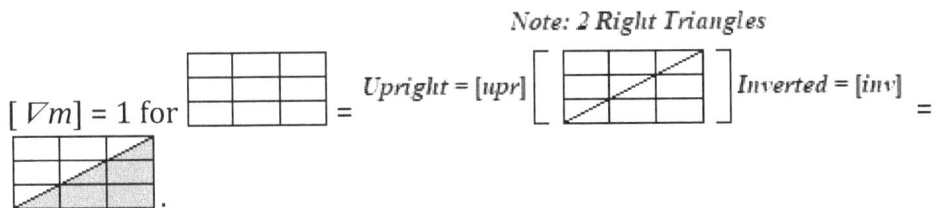

$[\nabla m] = 1$ for

$Upright = [upr]$ $Inverted = [inv]$

As opposed to...

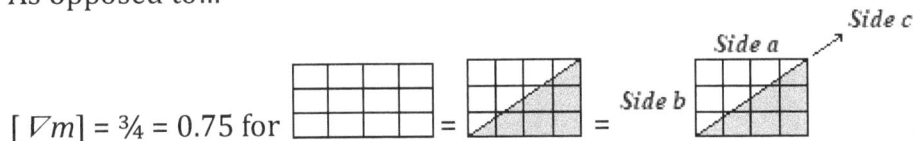

$[\nabla m] = \frac{3}{4} = 0.75$ for

Side a   *Side c*   *Side b*

That is ultimately equal to =

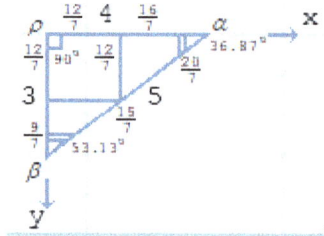

All of the 3-4-5-6 Golden Upright Right Triangle Measurement Calculations—The "3-4-5-6" [also known as the "4-3-5-6" and "3-5-4-6"] Golden Upright Right Triangle [or "GURT"]—In-Depth Measurements: Illustrating the transformation from the GURT back to the Visualus Isometric Cuboid as:

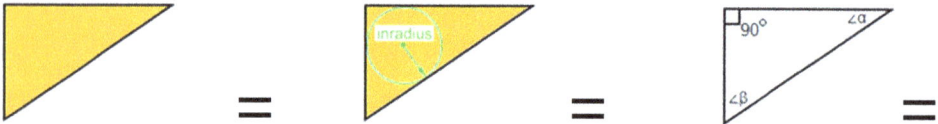

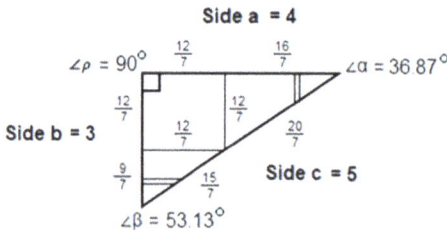

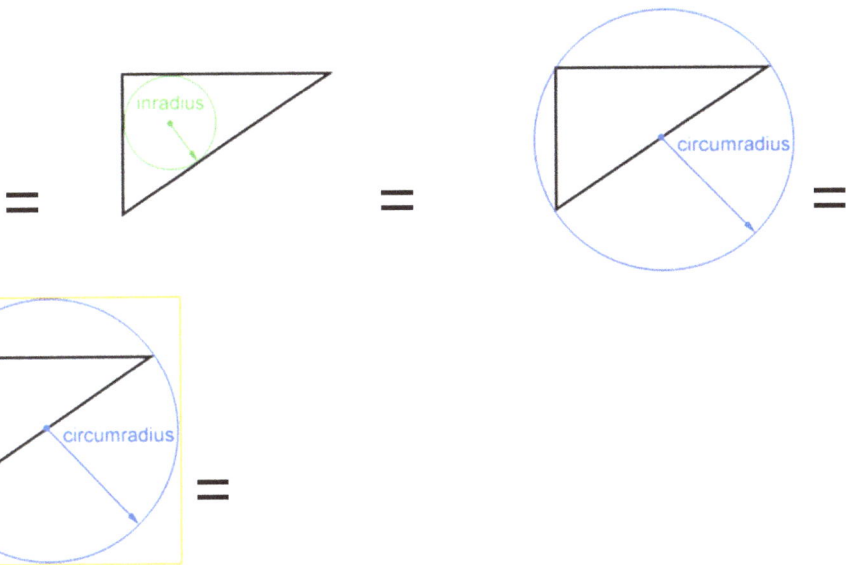

This then leads to the 3-4-5-6 Golden Upright Right Triangle—Semicircle Measurements (illustrating Thales Theorem)—that contain the Internal Right-Angled Inscribed Square (in the Golden Upright Right Triangle based on its 90-degree angle) by "Cubing the Square":

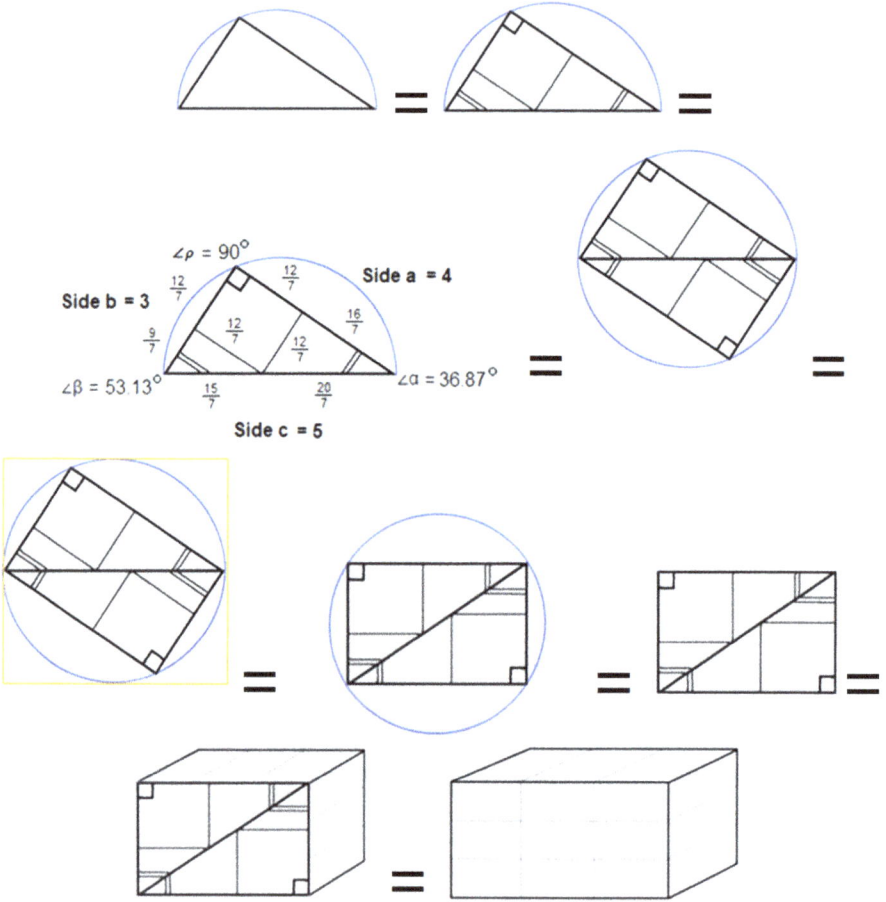

The 3-4-5 Golden Upright Right Triangle—Right Angled Internal Inscribed Square (in the Trichotomous Upright Right Triangle based on its 90-degree angle) with the External Circumradius That when sliced in half creates the Golden Upright Right Triangle Semicircle:

Science: The In-Depth Trichotomous Science of the Dynamic 3-4-5-6 Golden Upright Right Triangle for Innovative Problem-Solving. Osler Studios Incorporated ©, © Copyright 2022 All Rights Reserved.

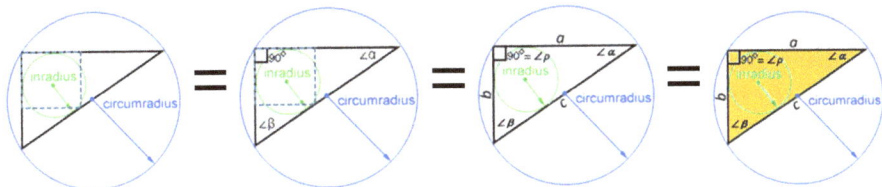

Overside Side c "Contraposition" Operations:

$$c = b[\Delta|x|] \equiv b\left[\sqrt{1 + \left[\frac{a}{b}\right]^2}\right]$$

Provides the length or "distance measure" of the Side c which is equal to the "Contraposition" of the Trichotomous Progression Line within the confines of the Upright Right Triangle which is one half of the Tri-Squared Analysis 3 by 3 Table Format and the front of the Visualis Isometric Cuboid - This particular equation and identification is referred to as the "Inclination Identification Equation".

The In-Depth descriptions of the Trichotomous Progression Line: Visualis Isometric Cuboid Front Side; the Triangle Equation Modeling Upright Right Triangle; and the Tri-Squared Analysis 3 by 3 Table.

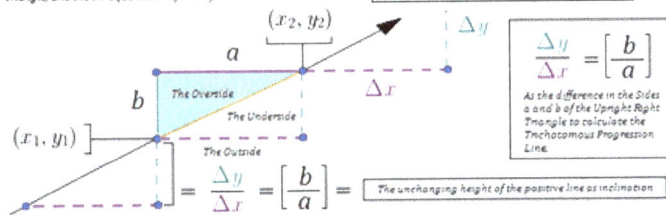

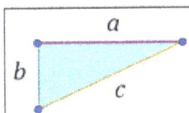

+ The positive direction of the Trichotomous Line

$$\frac{\Delta y}{\Delta x} = \left[\frac{b}{a}\right]$$

As the difference in the Sides a and b of the Upright Right Triangle relates to the Trichotomous Progression Line

$$= \frac{\Delta y}{\Delta x} = \left[\frac{b}{a}\right] = \text{The unchanging height of the positive line as inclination}$$

Overside Upright Right Triangle Measurements:

The Upright Right Triangle that is the Trichotomous Progression Analysis (or "TPA") "Trine". Note: a = 4; b = 3; and c = 5.

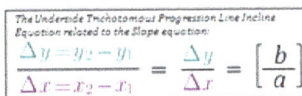

The Underside Trichotomous Progression Line Incline Equation related to the Slope equation:

$$\frac{\Delta y = y_2 - y_1}{\Delta x = x_2 - x_1} = \frac{\Delta y}{\Delta x} = \left[\frac{b}{a}\right]$$

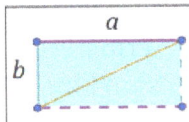

An illustration of the combined Overside and Underside displaying the combines Perimeter & Area:

The Tri-Square Test 3 by 3 Standard Table format and the "front face" or "front side" of the Visualis Isometric Cuboid.

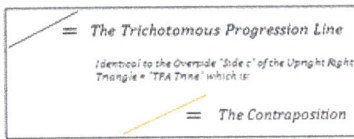

= The Trichotomous Progression Line

Identical to the Overside "Side c" of the Upright Right Triangle = "TPA Trine" which is:

= The Contraposition

The elongated Trichotomous Progression Line illustrated with the 3 by 3 Tri-Squared Analysis 3 by 3 Table creating the precise measurements for the Upright Right Triangle within the 3 by 3 Table creating the front of the Visualis Isometric Cuboid:

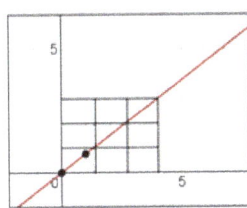

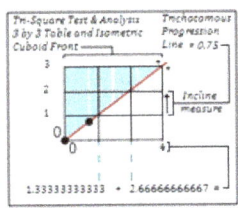

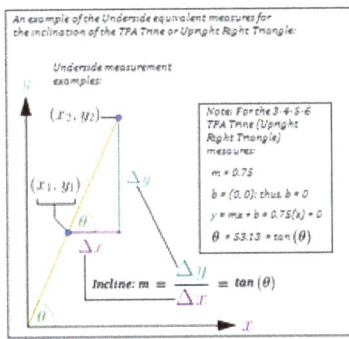

An example of the Underside equivalent measures for the inclination of the TPA Trine or Upright Right Triangle:

Underside measurement examples:

Tri-Square Test & Analysis 3 by 3 Table and Isometric Cuboid Front — Trichotomous Progression Line = 0.75

1.33333333333 ÷ 2.66666666667 =

Incline: $m = \frac{\Delta y}{\Delta x} = \tan(\theta)$

Note: For the 3·4·5·6 TPA Trine (Upright Right Triangle) measures:

m = 0.75
b = (0, 0); thus b = 0
y = mx + b = 0.75(x) + 0
θ = 53.13 = tan(θ)

$$\text{Incline: } m = \frac{\Delta y}{\Delta x} = \tan(\theta)$$

The Trioengineering Notation 3-4-5-6 Golden Upright Right Triangle—Right Angled Internal Inscribed Insquare and Inscribed Internal Incircle Inradius Complete Measurements

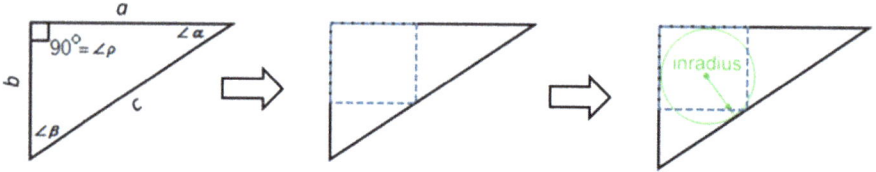

The Golden Upright Right Triangle Base Measurements = **3** (far left side) to the **4** (top side) to the **5** (diagonal side) or "∠*b* to ⬚*a for* ⎁**Side *c*"** = "***abc***" or "4-3-5-6" = "3-4-5-6" by the "Associative Property of Mathematics" which therefore creates the "3-4-5-6 Golden Upright Right Triangle" (moving by Side from Left to Top to Diagonal Right);

⎁**Side *a*** (top side indicating the *x*-axis) = 4 is analogous to the *x*-axis or abscissa line;

⎁**Side *b*** (far left side indicating the *y*-axis) = 3 is analogous to the *y*-axis or ordinate line;

⎁ **Side *c*** (diagonal side indicating the slope/inclination/acclivity) = analogous to the "G–TPA" (the acronym for the Triostatistics "Geometric Trichotomous Progression Analysis") Progression Line also the diagonal/hypotenuse = 5, that also the ⎁Right Triangle Incline and ⎁Right Triangle Acclivity;

Total Δ*abc* = ⎁*abc* = ⎁*abc* = 0.5 of a rectangle (or ½ of the Standard 3 by 3 Tri–Squared Test Table and the Front Face of the Visualus © Isometric Cuboid);

Total Δ*abc* in deference to [" ⎁"] = [" ⎁"] which represents the 3-4-5 Golden Upright Right Triangle as [⎁*abc*] = ⎁*abc* = 1;

∠***α*** = ⎁∠*α* = "Trichotomous Upright Right Triangle Angle alpha" = "Upper Far Right Angle" = 36.87° = 36°52'12" = 0.6435 radians (or "rad");

$\angle\beta$ = $\nabla\angle\beta$ = "Trichotomous Upright Right Triangle Angle beta" = "Lower Left Angle" = 53.13° = 53°7'48" = 0.9273 radians (or "rad");
$\angle\rho$ = $\nabla\angle\rho$ = "Trichotomous Upright Right Triangle Angle rho" = "Upper Right Angle" = 90° = 90°0'00" = 1.5708 radians (or "rad");
∇**Height "H" or "h"** = [h] = 2.4;
∇**Area "A"** = ∇A = 6;
∇**Perimeter "P"** = ∇P = 12;
∇**Inradius** also referred to by acronym in this case " ∇**Ird**"= 1;
∇**Circumradius** = 2.5; and
∇**Parallel Side Differential defined as "[Side *a* + Side *b* – Side *c*]"** or **[*a* + *b* – *c*]** = 2.
∇**Inscribed Insquare Definition** = The full internal expansion of the ninety-degree angle;
∇**Inscribed Insquare Side** = [12/7] = 1.714285714285714…;
∇**Inscribed Insquare One side [s]** = [12/7] = 1.714285714285714…;
∇**Inscribed Insquare Area [s²]** = $[12/7]^2$ = 2.938775510204082…; and
∇ **Inscribed Insquare Perimeter [4(s)]** = 4[12/7] = 4(1.714285714285714…) = [12/7] + [12/7] + [12/7] + [12/7] = 6.857142857142857…

Trioengineering Notation Defining and Explaining the Rigorous, Meticulous, and Precise Trichotomization Mathematical Operation

The Final "Trioengineering Trichotomization Equation":

$$\overset{3}{\underset{i=1}{\nabla}}\left[\, \text{Ʒ}\,\right] = \nabla y = y_1 + [\nabla x - x_1]\frac{(y_2 - y_1)}{(x_2 - x_1)} = \nabla\,[\text{TEM}]_{[\text{M.I.}]}.$$

How the "Trioengineering Trichotomization Equation" operates:

$$\overset{3}{\underset{i=1}{\nabla}}[\mathbf{3}] = \nabla y = y_1 + [\nabla x - x_1]\frac{(y_2 - y_1)}{(x_2 - x_1)} = \nabla[\text{TEM}]_{[M.I.]} = P_2 \underset{c}{\overset{b}{\begin{vmatrix}[\text{M.I.}]\end{vmatrix}}}\overset{P_1}{\underset{P_3}{\longleftarrow}}a .$$

		Main Idea: "…" = $\nabla[P_1 + P_2 + P_3] = I_3$			
***abc* Categorization Classification**	**Mathematical Law of Trichotomy as Triological Definitions**	**Part of the Main Idea**	**Equal to →**	**Name: "…"**	**Area of ∇abc on the [TEM**
a	$a \neg b \wedge c$ *Defined as:* "a not b nor c"	∇P_1	=	"Name" of ∇P_1	$\nabla a =$ "Side a"
b	$b \neg a \wedge c$ *Defined as:* "b not a nor c"	∇P_2	=	"Name" of ∇P_2	$\nabla b =$ "Side b"
c	$c \sim a \wedge b$ *Defined as:* "c neither a nor b"	∇P_3	=	"Name" of ∇P_3	$\nabla c =$ "Side c"

The Trichotomization Table for "[Tri]*abc*" as an In-Depth Categorization Classification Analysis for the Meticulous and Precise Creation of a Triangular Equation Model [TEM]

The Trichotomization Table requires the solution-seeker to "think in threes". Mor specifically, this mode of thinking is defined and explained in detail in the Table rows an columns that appear above. The "Main Idea" is viewed as three separate parts that ar interdependent and creat the "Main Idea" as a whole. This is the very essence of th "Mathematical Law of Trichotomy" and is illustrated in the second colum as "Triologic". base example of this is give in the Triological Science of "Triology" in which the atom is th base and basic subatomic structure and primary infrastructure of all matter. The atom trichotomously composed of the: proton, neutron, and the electron. This sam trichotomous structure can be applied to the "Main Idea" in the center of the [TEM] an parts 1 though 3 (as "P1, P2, P3" respectively) compose and complete the outside of th Triangular Equation Model [TEM] = "I_3" which is the trichotomous "Interpolation" (that indicative of the "exterior" of the GURT) of Trioengineering Notation.

Chapter Three follows and further defines Trioengineering.

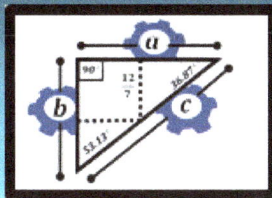

And he said, Hear now my words: If there be a prophet among you, I the LORD will make myself known unto him in a vision, and will speak unto him in a dream.

Numbers 12: 6

A Detailed Explanation of How Trioengineering Works via In-Depth Metagraphic Mathematical Models

Introduction to the Trioengineering Methodology

The field of Trioengineering studies the solutions to various problems using the framework of "Structural Standards" grounded in the universal "Mathematical Law of Trichotomy". The universal model is explained and defined in the scientific field of Triology. Triology starts with the base universal structure of all matter (which is the trichotomous structure of the atom) that is explored in the Triological Science scientific field of Triophysics). In Trioengineering the base form of problem-solving as the foundation of all of the Triological Sciences via the 3-4-5-6 Golden Upright Right Triangle (abbreviated as the GURT). The 3-4-5-6 GURT is used to create the solution providing instrumentation and models via the scientific fields of Triomathematics, Ternary Algebra, and Triostatistics (via the Tri–Squared Test and its unique instrumentation used for research and problem-solving). The Triostatistics Tri–Squared Test is the basis and basic starting point of all Triostatistical operations and research methodologies. The Triostatistical methodology of Triangular Equation Modeling [TEM] that together provide the standard structural model for problem-solving analysis. The Trioengineering Ecosystem as a problem-solving model has the following "***Trioengineering Standard Structures***":

1.) The [TEM] (from Triostatistics);

2.) The Tri^2 3 by 3 initial structural Table format to form the initial 3-4-5-6 GURT; and lastly

3.) Unique 3-4-5-6 GURT Metacognetic Mechanics Metagraphics (from Visualus).

Trioengineering Notation in Action: Trioengineering Notation Defined

$$\nabla \text{ or } \nabla N = \sum_{i=1}^{3} \text{ for } \left.\begin{array}{l} a = 4; \\ b = 3; \\ c = 5; \text{ and} \\ A = 6. \end{array}\right\} = \nabla abc = \quad \text{as [TEM]} =$$

expressed also as the Tripositive = ∇ = [Tri].

Where,

N = "Name" of the measurement/concept/solution; and
∇ = "The 3-4-5-6 Golden Upright Right Triangle" or "Triune" in action as the "Trioengineering Trichotomate" and " Trichotomous Triangulation" = "Trioengineering Trichotomate".

Thus,

$$\nabla N = \sum_{i=1}^{3} [N] \equiv \nabla Name_1 + \nabla Name_2 + \nabla Name_3 = [\nabla Name_1; \nabla Name_2;$$

and $\nabla Name_3$] for Primary Main Name in:

∇I_1 = "Trichotomous Triangle Inculcation" ≡ , ∇M = Middle of $\nabla[M] = \nabla[\mathbf{3}]$;

∇I_2 = "Trichotomous Triangle Intercalation" ≡ $\nabla[TEM]$; and

∇I_3 = "Trichotomous Triangle Interpolation" ≡ $\nabla[$"Side c"].

In addition, the following applies:

∇... = "The Trioengineering Trichotomation of: "..." "; (Note: the term "Trichotomation" is defined as the separation into units of 3 based upon "The Mathematical Law of Trichotomy") and

∇N = "The Trioengineering Trichotomation of: "N" "; with the following mathematical definitions that are applicable to the aforementioned trichotomy—

∇Inculcation$_1$ = " ∇Interior" = Inside of the 3-4-5-6 GURT as the [M.I.] = [Main Idea] = $\nabla[\mathbf{3}]$ = The Overall Overriding Subject;

∇Intercalation$_2$ = " ∇Exterior" = "Trioengineering Trichotomation" = ∇ [TEM] ;

∇Interpolation$_3$ = " ∇ Ulterior" = The Trigmoid Function for "Side c" of the 3-4-5-6 GURT = The Final Solution as Side c = "*convey*".

Note the following:

$$\nabla[\mathbf{3}] = \nabla y = y_1 + [\nabla x - x_1]\frac{(y_2 - y_1)}{(x_2 - x_1)};$$

$$\nabla[\mathbf{3}] = \nabla y = 0 + [3 - 0]\frac{(3 - 0)}{(3 - 0)};$$

$$\nabla[\mathbf{3}] = \nabla y = 3$$

$$V = \overset{3}{\underset{i=1}{V}} = [\text{Tri}]$$ as "Trioengineering Trichotomous Triangulation" =

"Trichotomation" = The combination of $I_1...I_3...$ This is further defined with the following 4 Primary Cartesian Coordinates which are $\{(x_1 = 0, x_2 = 3); (y_1 = 0, y_2 = 3); (x_1 = 0, y_1 = 0);$ and $(x_2 = 3, y_2 = 3;)$ as $V[3:3]$ with the same Cartesian Coordinates except for the last for $V[4:3]$ that are $(x_3 = 4, y_3 = 3;)\}$ and mathematically in the following manner:

$$\overset{3}{\underset{i=1}{V}}[\mathbf{3}] = Vy = y_1 + [Vx - x_1]\frac{(y_2 - y_1)}{(x_2 - x_1)} = V[\text{TEM}] = [\text{TEM}], \quad \text{for}$$

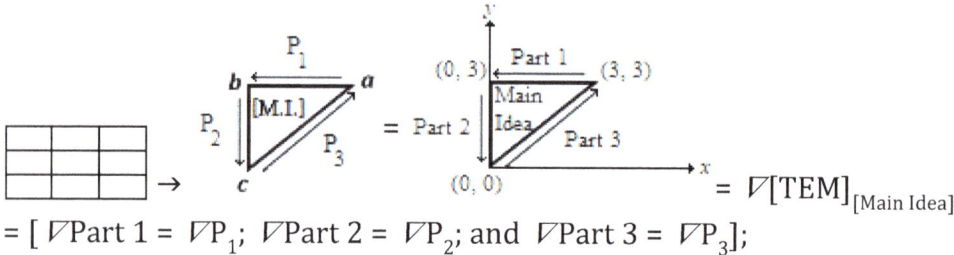

$$= [\ V\text{Part 1} = VP_1; \ V\text{Part 2} = VP_2; \text{ and } V\text{Part 3} = VP_3];$$

$$\overset{3}{\underset{i=1}{V}}[\mathbf{3}] = \text{Part 3} = (\text{Main Idea} - \text{Part 1} + \text{Part 2});$$

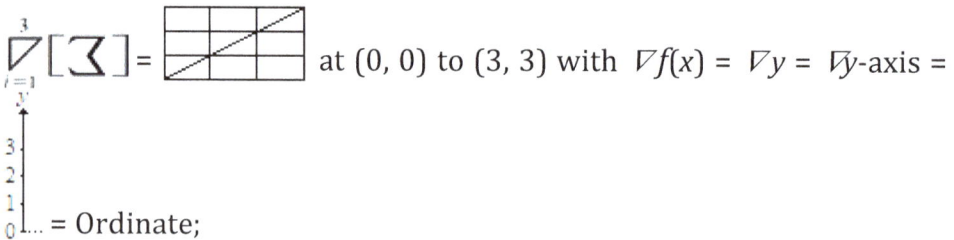

$$\overset{3}{\underset{i=1}{V}}[\mathbf{3}] = $$ at (0, 0) to (3, 3) with $Vf(x) = Vy = Vy\text{-axis} =$

... = Ordinate;

 $$= Vy = 0 + [3 - 0]\frac{(3 - 0)}{(3 - 0)} = 3(3/3) = 3(1) = 3;$$

$$\nabla y = 3 = \nabla[\text{TEM}] =$$

$$\to \nabla[\text{M. I.}];$$

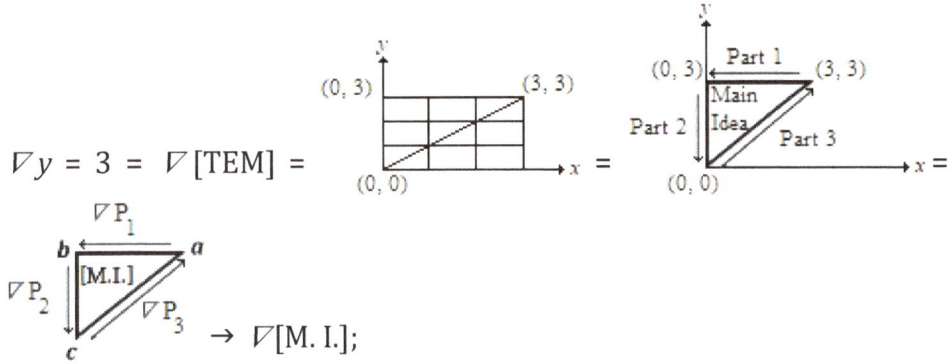

From rectangle to Square: The Coequality of [3:3] with [4:3] is graphically expressed as:

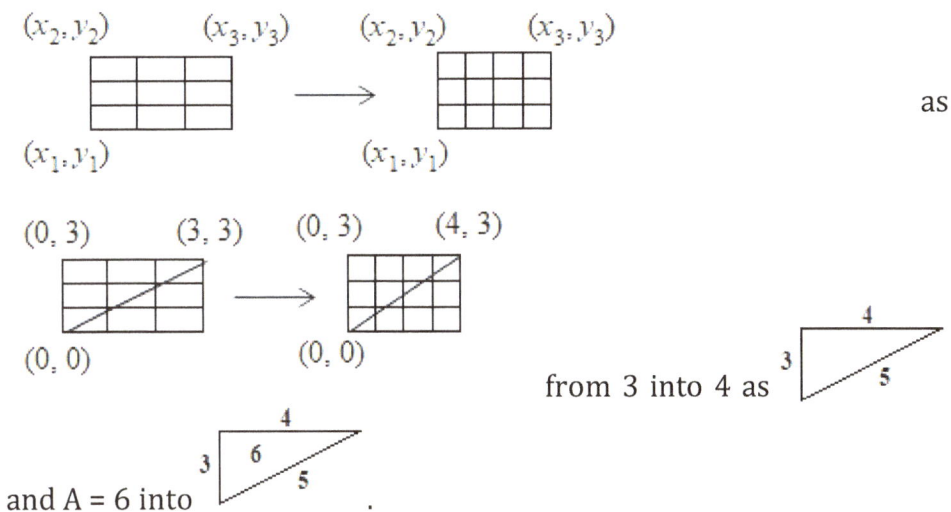

as

from 3 into 4 as

and A = 6 into

.

Where,

$$\nabla_{[3:3] \text{ and } [4:3]} \equiv (x_0, y_0) = (0, 3);$$
$$\nabla_{[3:3] \text{ and } [4:3]} \equiv (x_1, y_1) = (0, 0);$$

$V_{[3:3]} \equiv (x_2, y_2) = (3, 3)$; and
$V_{[4:3]} \equiv (x_3, y_3) = (4, 3)$.

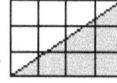

In terms of "Trichotomous Conversion" for = [3](4/3) = 4/1 = 4 for Side *a*.

Note: For $V\left[\mathbf{3}\right]$ = " " = "The Trioengineering Tripositive Trigmoid"…

Thus,

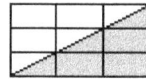

For the 3 by 3 Table that has the $V_{[3:3]}$ = " ", Note the following:

$$V[\mathbf{3}] = \overset{3}{\underset{i=1}{V}}[\mathbf{3}] = Vy = y_1 + [Vx - x_1]\frac{(y_2 - y_1)}{(x_2 - x_1)}$$, where, (x_1, y_1) and (x_2, y_2) = (0, 0) and (3, 3) respectively, with the Coequality Conversion Equation for the 3 by 3 Table = [3:3] into the 4 by 3 Table = [4:3] for the " " as:

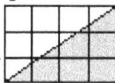

$$Vx_{[4:3]} = y_1 + [Vy - y_1]\frac{(y_3 - y_1)}{(x_3 - x_1)}$$, and [4 by 3] for the equation [4:3] Cartesian Coordinates where [4 by 3] = [4:3] has (x_1, y_1) and (x_3, y_3) respectively = (0, 0) and (4, 3). For the 3 by 3 Table conversely, $Vx_{[3 \text{ by }3] = [3:3]}$ = 3 for " ".

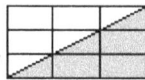

It is important to note that the following is therefore true:

" " = 4 by 3 Table = 4:3 Table Ratio (indicating the Trioengineered 3-4-5-6 GURT) = ∇ [4:3] = The Trioengineered Tripositive 4 by 3; and

" " = 3 by 3 Table = 3:3 Table Ratio (indicating the Trioengineered 3-4-5-6 GURT) = ∇ [3:3] = The Trioengineered Tripositive 3 by 3.

Furthermore, the following is true and applies to the aforementioned:

" " = ∇[4:3] = The Trioengineered GURT in the 4 by 3 Table = $\nabla x_{[4:3]}$ = 0 + [3 − 0](4 − 0/3 − 0) = [3](4/3) = (4/1) = 4; and

" " = ∇[3:3] = 3:3 The Trioengineered GURT in the 3 by 3 Table = $\nabla x_{[3:3]}$ = 3.

The Trioengineering Tripositive Trigmoid Identities

$\nabla a_{[4:3]}$ ≡ The Trioengineering Tripositive of "Side a" on the 4 by 3 (or [4:3]) Table GURT in the following manner:

$$\nabla[\text{3}] = \overset{3}{\underset{i=1}{\nabla}}[\text{3}] = \nabla x = y_1 + [\nabla y - y_1]\frac{(y_3 - y_1)}{(x_3 - x_1)}.$$

The abovementioned mathematical geometric formula justifies "Side a" for the [4:3] in a graphical fashion.

$\nabla b_{[3:3] \text{ and } [4:3]} \equiv$ The Trioengineering Tripositive of "Side b" on the 3 by 3 (or [3:3]) Table GURT and 4 by 3 (or [4:3]) Table GURT respectively in the following manner:

$$\nabla[3] = \nabla_{i=1}^{3}[3] = \nabla y = y_1 + [\nabla x - x_1]\frac{(y_2 - y_1)}{(x_2 - x_1)}.$$

The abovementioned mathematical geometric formula justifies "Side b" for the [3:3] and [4:3] in a graphical fashion.

$\nabla c_{[3:3]} \equiv$ The Trioengineering Tripositive of "Side c" on the 3 by 3 (or [3:3]) Table GURT and 4 by 3 (or [4:3]) Table GURT respectively in the following manner:

$$\nabla[3] = \nabla_{i=1}^{3}[3] = \nabla c_{[x=y] \equiv [3:3] \equiv [4:3]} = \left[\nabla b \sqrt{1 + \left[\frac{\nabla a}{\nabla b}\right]^2} \right].$$

The abovementioned mathematical geometric formula justifies "Side c" for [3:3] and [4:3] in a graphical fashion.

Chapter Four follows and further defines the Trioengineering Trigmoid.

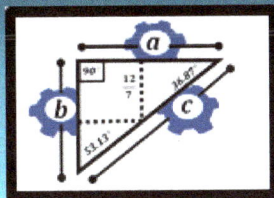

"Where there is no vision, the people perish: but he that keepeth the law, happy is he."

Proverbs 29: 18

The Trigmoid: Symbol, Curve, and Function

Explaining the Trigmoid in Implicit and Explicit Detail via Mathematical Equations, Geometric Models, In-Depth Graphics and how they all work together using the In-Depth precision, meticulousness, and rigor of Explicative Trioengineering Notation. The Sigmoid curve illustrated below provides an example of exactly how the "Trigmoid" is designed.

The Base Sigmoid Symbol, Curve, and Function

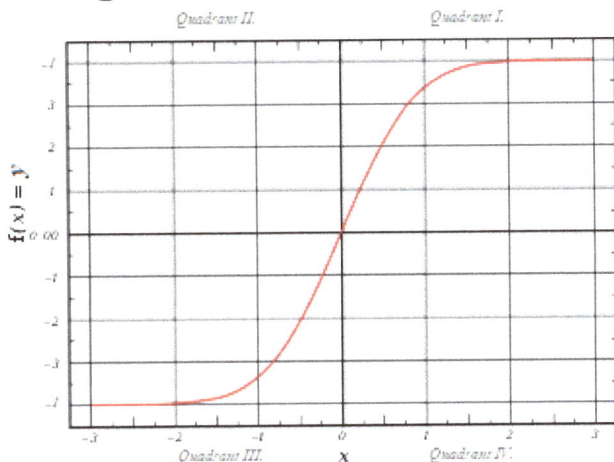

Trigmoid

" З "

Defining and Describing the "Trigmoid" Symbol

The "Trigmoid" symbol is an inverse or "reversed" Capital Sigma that becomes a an angular "3". In fact, this is by design as the Trigmoid is meant to represent "3" as the "Triune Trichotomous Trigmoid" defines the 3-4-5-6 GURT in terms of the symbolism, angularity, and functionality of the scientific field of Trioengineering.

The Trigmoid (Math) Symbol

The Trigmoid Curve (Line)

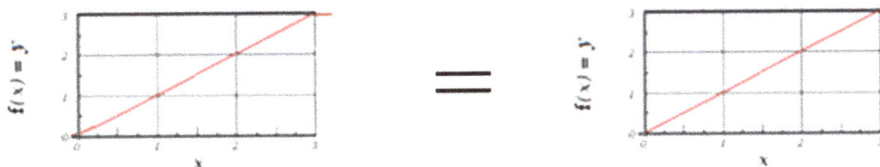

The Trigmoid Function (Shape)

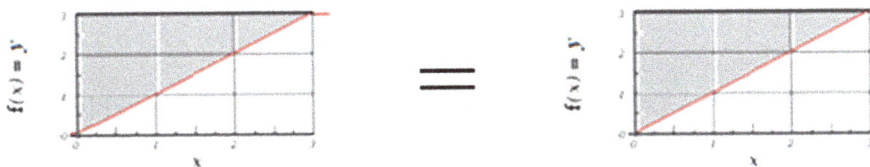

The "Trigmoid" Symbol: The "Trigmoid Symbol" represents both the "Trigmoid Function" and the "Trigmoid Curve". The symbol is a reverse Greek capital letter "Sigma" ("Σ") that resembles (when reversed) an angular number "3". In fact, the Trigmoid Symbol represents the "Trichotomous Number 3" as it applies to all of the aspects of the "3-4-

5-6 Golden Upright Right Triangle" in terms of its initial extraction from the Visualus Isometric Cuboid and its direct application to the 3 by 3 Table (that is the inherent part of the Tri–Squared Test) in terms of its intrinsic: angles (that trichotomously number 3), sides (that also trichotomously number 3), and its application in the field of "Triostatistics" as the "Triangular Equation Modeling" = [TEM] method of analysis. [TEM] uses the "3-4-5-6 Golden Upright Right Triangle" in terms of its center and sides (that also trichotomously number 3—as the primary and critical three categories, concepts and/or ideas, or prototype, build and/or construction, or designs, innovations, and/or solutions that are presented geometrically and graphically). The Trigmoid Symbol is used in the "Trichotomous Triangular Trigmoid Equation" = Side c in the following Trioengineering Notation (that uses the "Trine Symbol" " ∇ ") equation:

$$\nabla y = y_1 + (\nabla x - x_1)\frac{(y_2 - y_1)}{(x_2 - x_1)} = \overset{3}{\underset{i=1}{\nabla}} [\Xi]$$

into

$$\nabla y = y_1 + (\nabla x - x_1)\frac{(y_2 - y_1)}{(x_2 - x_1)} = \overset{3}{\underset{i=1}{\nabla}} [\Xi]$$

Under the following conditions:
[The 3 by 3 Table with $m = 1$]: (0, 0) to (3, 3) is immutable.

"Trigmoid Curve" Definition: A "Trigmoid Curve" is an elongated Sigmoid Curve that has been stretched to conform to Side c on the " ∇abc " Upright Right Triangle that fits on the 3 by 3 Table of the Tri–Squared Test (that is also the "Front Face" of the Visualus Isometric Cuboid). There is no negative continuation nor positive continuation of the "Trigmoid Curve" due to its perfect fit of Side c of the trichotomous

Upright Right Triangle, where, Side c begins at (0, 0) and ends at (4, 3). The Side c Equation using "Trioengineering Notation" (the notation that represents the use of the Trigmoid symbol to mathematically define and explain the unique nature and relationship of trichotomy within the $\sqrt{abc}$ to create Side c in a graphical representation). The "Trigmoid Curve" therefore becomes the "Trigmoid Line" = Side c. The Side c Trioengineering Notation Trigmoid Equation (which leads to the geometric illustration) is written in the following manner:

$$\overset{3}{\underset{i=1}{\triangledown}} [3] = $$

"Trigmoid Function" Definition: The Trigmoid Function justifies the Side c line that is a trichotomous part of the 3-4-5-6 Golden Upright Right triangle. A "Trigmoid Function" unilateral way of defining a shape that is a "trichotomously bounded function" (in this instance a "trichotomously bounded function" is where there exists a quantifiable a real number n[x1...x5] such that for all positive integers x are in X, where X = m as the Side c slope/incline/acclivity of "$\sqrt{abc}$"). Additionally, in this instance the Trigmoid Function is "trichotomously differentiable" (where "trichotomously differentiable" is defined as becoming 'different' via acclivity or incline illustrating the process of geometric growth and development). The Trigmoid Function is a real function that is defined for all real input values and has a non-negative derivative at each point and no inflection point(s). A Trigmoid "Function" and a Trigmoid "Curve" refer to the same object (with a m = 1 on the 3 by 3 Table and transforms into m = 0.75 on the 4 by 3 Table).

In Trioengineering Notation this is written mathematically, geometrically, and graphically on the 3 by 3 Table is as follows:

$$\overset{3}{\underset{i=1}{\nabla}}[\textrm{Ʒ}] = \ = \ = \ =$$

Trigmoid

$$``\textrm{Ʒ}" \quad = \quad = $$

The Trichotomous Triangular Trigmoid $= \overset{3}{\underset{i=1}{\nabla}}[\textrm{Ʒ}]$

$$\overset{3}{\underset{i=1}{\nabla}}[\textrm{Ʒ}] = \ = \ = $$

$$= \ = $$

Trichotomous Three $= \textrm{Ʒ} = 3$

There is truly a trichotomy in the aspects and characteristics of the Trigmoid that is: 1.) A "function" (or shape); a "curve" (that conforms to a line with incline/acclivity); and ultimately a line (that becomes the side of the 3-4-5-6 GURT). Furthermore, there is a trichotomy of lines within the Trigmoid function or shape that creates the final the 3-4-5-6 GURT as sides: 1.) ∇a; 2.) ∇b; and lastly ∇c respectively.

Trigmoid

"3" = =

The Sigmoid Symbol, Curve, and Function

That is truly Converted, is ultimately Transformed, and finally Conforms into the following as the Trigmoid that illustrates "Side c" in Trioengineering Notation:

$$\overset{3}{\underset{i=1}{\nabla}}[\mathbf{3}] = \frac{\nabla abc}{\nabla a \nabla b} = \frac{\nabla[abc]}{\nabla[ab]} = \nabla c.$$

Thus,

$$Vc_{V[x=y] \equiv V[3:3] \equiv V[4:3]} \equiv \overset{3}{\underset{i=1}{\nabla}}_{[Line Vc]}[\mathbf{3}] = \left[Vb \sqrt{1 + \left[\frac{Va}{Vb}\right]^2}\right] \quad = \quad \text{[graph]} \quad = \quad \overset{3}{\underset{i=1}{\nabla}}[\mathbf{3}] = \frac{\nabla abc}{\nabla a \nabla b} = \frac{\nabla[abc]}{\nabla[ab]} =$$

The Trigmoid Equation

$$\overset{3}{\underset{i=1}{\nabla}}[\mathbf{3}] = $$

The Trigmoid Equation, illustrates the equity and thought contained within the Trigmoid Model. This is shown as a problem-solving and solutions-driven model that is produced and illustrated graphically as a "Triangular Equation Model". The entire Trigmoid Equation is presented in Trioengineering Notation in the following manner:

$$\nabla[TEM]_{[Idve]} = \quad \text{[triangle diagram]} \quad \longrightarrow \quad \nabla[TEM]_{[MI]} = \quad \text{[triangle diagram]}$$

The three trichotomous Trigmoid Equations and Identities, that create the 3-4-5-6 GURT thereby illustrating the equity and thought contained within the Trigmoid Model and is universally applicable to both the 3 by 3 Table ratio and the 4 by 3 Table ratio respectively. They are:

$$\nabla b \,_{\nabla[3:3]} \equiv \nabla[4:3] \equiv \sum_{i=1}^{3} [\textbf{3}]_{[Line\ Vb]} = \nabla y = y_1 + [\nabla x - x_1]\frac{(y_2 - y_1)}{(x_2 - x_1)};$$

$$\nabla a \,_{\nabla[4:3]} \equiv \nabla x \,_{\nabla[4:3]} \equiv \sum_{i=1}^{3} [\textbf{3}]_{[Line\ Va]} = y_1 + [\nabla y - y_1]\frac{(y_3 - y_1)}{(x_3 - x_1)}; \text{ and}$$

$$\nabla c \,_{\nabla[x=y]} \equiv \nabla[3:3] \equiv \nabla[4:3] \equiv \sum_{i=1}^{3} [\textbf{3}]_{[Line\ Vc]} = \left[\nabla b \sqrt{1 + \left[\frac{\nabla a}{\nabla b}\right]^2} \right].$$

Note: the following in regards to the three trichotomous Trigmoid Equations and Identities, that parsimoniously create the 3-4-5-6 GURT thereby illustrating the equity and thought contained within the Trigmoid Model and is universally applicable to both the 3 by 3 Table ratio and the 4 by 3 Table ratio that can now be simplified and expressed in terms of the Trioengineered capital Greek letter Delta (Δ) that is indicative of the change in *y* over *x* Cartesian Coordinates for the Trioengineered Sides *b* and *a* respectively. They are represented mathematically as follows:

$$\frac{\nabla[\Delta y_2]}{\nabla[\Delta x_2]} = \frac{(y_2 - y_1)}{(x_2 - x_1)}; \text{ and}$$

$$\frac{\nabla[\Delta y_3]}{\nabla[\Delta x_3]} = \frac{(y_3 - y_1)}{(x_3 - x_1)}.$$

The three trichotomous Trigmoid Equations and Identities, that create the 3-4-5-6 GURT thereby illustrating the equity and thought contained within the Trigmoid Model and is universally applicable to both the 3 by 3 Table ratio and the 4 by 3 Table ratio can be expressed using the Trioengineered capital Greek letter Delta (Δ) that is indicative of the change in *y* over *x* Cartesian Coordinates respectively. This represented mathematically in the following formulae:

$$\nabla b_{\nabla[3:3]\,\equiv\,\nabla[4:3]} \equiv \nabla\left[\underset{[Line\ \nabla b]}{\boldsymbol{\zeta}}\right] = \nabla y = y_1 + [\,\nabla x - x_1\,]\frac{\nabla[\Delta y_2]}{\nabla[\Delta x_2]};$$

$$\nabla a_{\nabla[4:3]} \equiv \nabla x_{\nabla[4:3]} \equiv \nabla\left[\underset{[Line\ \nabla a]}{\boldsymbol{\zeta}}\right] = y_1 + [\,\nabla y - y_1\,]\frac{\nabla[\Delta y_3]}{\nabla[\Delta x_3]};\ and$$

$$\nabla c_{\nabla[x\,=\,y]\,\equiv\,\nabla[3:3]\,\equiv\,\nabla[4:3]} \equiv \nabla\left[\underset{[Line\ \nabla c]}{\boldsymbol{\zeta}}\right] = \left[\nabla b\,\sqrt{1 + \left[\frac{\nabla a}{\nabla b}\right]^2}\right].$$

As such the following is true for the equations and identities associated with "Trioengineered Side b" [" ∇b"] as the "Total Trioengineering [Side b] Triangulation Equations and Identities:

$$\nabla b_{\nabla[3:3]\,\equiv\,\nabla[4:3]} \equiv \nabla\left[\underset{[Line\ \nabla b]}{\boldsymbol{\zeta}}\right] = \nabla y = y_1 + [\,\nabla x - x_1\,]\frac{(y_2 - y_1)}{(x_2 - x_1)}$$

Is identical to:

$$\nabla b_{\nabla[3:3]\,\equiv\,\nabla[4:3]} \equiv \nabla\left[\underset{[Line\ \nabla b]}{\boldsymbol{\zeta}}\right] = \nabla y = y_1 + [\,\nabla x - x_1\,]\frac{\nabla[\Delta y_2]}{\nabla[\Delta x_2]}.$$

As such the following is true for the equations and identities associated with "Trioengineered Side a" [" ∇a"] as the "Total Trioengineering [Side a] Triangulation Equations and Identities:

$$Va_{\nabla[4:3]} \equiv \nabla x_{\nabla[4:3]} \equiv \nabla_{i=1}^{3}\left[\underset{[Line\ Va]}{\sum}\right] = y_1 + [\nabla y - y_1]\frac{(y_3 - y_1)}{(x_3 - x_1)}$$

s identical to:

$$Va_{\nabla[4:3]} \equiv \nabla x_{\nabla[4:3]} \equiv \nabla_{i=1}^{3}\left[\underset{[Line\ Va]}{\sum}\right] = y_1 + [\nabla y - y_1]\frac{\nabla[\Delta y_3]}{\nabla[\Delta x_3]}.$$

$$Vb_{\nabla[3:3] \equiv \nabla[4:3]} \equiv \nabla_{i=1}^{3}\left[\underset{[Line\ Vb]}{\sum}\right] = \nabla y = y_1 + [\nabla x - x_1]\frac{(y_2 - y_1)}{(x_2 - x_1)}$$

The abovementioned is the detailed equation and identity of ∇b. It is literally defined in the following manner: "Trioengineered Side b is applied to the Trioengineered 3 by 3 Table ratio that is logically equivalent to the Trioengineered 4 by 3 Table this is identical to the Trioengineered Trigmoid for Trioengineered Line b that is equal to the Trioengineered y equal to y sub-one plus Trioengineered x minus x sub-one times the slope/incline/acclivity of the first two Trigmoid Cartesian Coordinates y sub-two minus y sub-one over x sub-two minus x sub-one".

$$\nabla b\,_{\nabla[3:3]} \equiv \nabla[4:3] \equiv \overset{3}{\underset{i=1}{\nabla}}\underset{[Line\ Vb]}{\left[\,\mathbf{3}\,\right]} = \nabla y = y_1 + [\,\nabla x - x_1\,]\frac{\nabla[\Delta y_2]}{\nabla[\Delta x_2]}$$

The abovementioned is the detailed "Total Trioengineering [Side b] Triangulation Equations and Identities for ∇b. It is literally defined in the following manner: "Trioengineered Side b is applied to the Trioengineered 3 by 3 Table ratio that is logically equivalent to the Trioengineered 4 by 3 Table this is identical to Trioengineered Trigmoid for Trioengineered Line b that is equal to the Trioengineered y equal to y sub-one plus Trioengineered x minus x sub-one times the slope/incline/acclivity of the first two Trigmoid Cartesian Coordinates that is Trioengineered Delta y sub-two over Trioengineered Delta x sub-two".

$$\nabla a\,_{\nabla[4:3]} \equiv \nabla x\,_{\nabla[4:3]} \equiv \overset{3}{\underset{i=1}{\nabla}}\underset{[Line\ Va]}{\left[\,\mathbf{3}\,\right]} = y_1 + [\,\nabla y - y_1\,]\frac{(y_3 - y_1)}{(x_3 - x_1)}$$

The abovementioned is the detailed equation and identity of ∇a. It is literally defined in the following manner: "Trioengineered Side a is applied to the Trioengineered 4 by 3 Table ratio that is identical to the Trioengineered x applied to the Trioengineered 4 by 3 Table this is identical to Trioengineered Trigmoid for Trioengineered Line a that is equal to y sub-one plus Trioengineered y minus y sub-one times the slope/incline/acclivity of the first third Trigmoid Cartesian Coordinate y sub-three minus y sub-one over x sub-three minus x sub-one".

$$Vc \, _{V[x=y] \equiv V[3:3] \equiv V[4:3]} \equiv \overset{3}{\underset{i=1}{\bigvee}} \underset{[Line \ Vc]}{[\mathbf{3}]} = \left[Vb \sqrt{1 + \left[\frac{Va}{Vb} \right]^2} \right]^2$$

The abovementioned is the detailed equation and identity of Vc. It is literally defined in the following manner: "Trioengineered Side c is applied to the Trioengineered Cartesian Coordinates x equal to y that is logically equivalent to the Trioengineered 3 by 3 Table ratio that is logically equivalent to the Same Side on the Trioengineered 4 by 3 Table ratio this is identical to the Side of the Trioengineered Trigmoid (with 3 trichotomous sides at an index of 1 and ending in 3 Sides) for Trioengineered Line c that is equal to the concentration on Side b times the square root of 1 plus the square of the slope/incline/acclivity of the 4 by 3 Table ratio that is Trioengineered a over Trioengineered b".

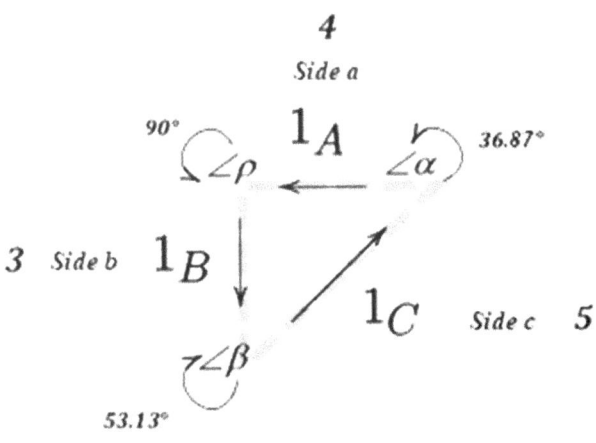

The abovementioned is the detailed model of the Triological Science Triostatistics "Triangular Equation Model" or [TEM]. The trichotomy of all three sides is intact as the "Trichotomous Morphisms" (the cyclical arrows) move to each adjacent side creating a unique trichotomous whole that the [TEM] then presents as the "Main Idea that appears in the exact center of the model. Although individual and unique in their own way each of the Trioengineered trichotomous sides is equal, thus:

$$\text{``} \, \nabla \, [1A = 1B = 1C]\text{''}.$$

Under the following conditions: [The 3 by 3 Table with $m = 1$]: $(0, 0)$ to $(3, 3)$ is immutable.

Chapter Five follows and explains in detail how to calculate with Trioengineering.

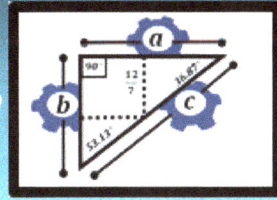

Wait for it; because it will surely come, it will not tarry.

Habakkuk 2: 3

Develop an In-Depth Comprehension of the GURT to Gain a Greater Understanding of the Trigmoid Function

The next sections provide more in-depth insight into the use of the Trigmoid in the scientific field of Trioengineering.

Trioengineering Thought or Thinking in Terms of Application and Learning

The application of the 3-4-5-6 GURT in terms of: "How to think and operate in divinely inspired threes and to research and decision-make in terms and via "The Mathematical Law of Trichotomy". The aforementioned occurs in the following manner:

1.) The manifestation of the initial thought; idea; or concept;
2.) The break down of the initial thought into the a.) primary idea or thought; b.) the establishment of the oppositional idea or thought; and the final lastly the production of the neutral (or non-a. and non-b.) idea, thought, or concept.

Triflow © which defined as the portmanteau of the two dual terms of "Trichotomy" in terms of "Trichotomous Thinking" combined with the term "Workflow". "Trichotomous Thinking" (in terms of the application of the 3-4-5-6 GURT and [TEM]) that is the separation into threes using

---

the 3-4-5-6 GURT is smooth from a holistic perspective. In this instance it is centered around a central "theme", "idea", and/or "concept". It is then trichotomously divided into its 3 main parts that are then termed as "sectors" = "parameter or sector *a*" that is opposed by "parameter or sector *b*" and lastly presents "parameter or sector *c*" which is neither "*a* or *b*" (adapted and revised from Osler, 2019—and the research conducted in the published research articles from Osler 2020-21).

Trioengineering Tri–Point Triangulation

Trioengineering Tri–Point Triangulation is a methodology that carefully and completely explains the use of the "Mathematical Law of Trichotomy" to define and model and define the [TEM] in terms of Trioengineering angular data points that are termed as: "Trichotomous Tri–Points 1 through 3".

The Trioengineering Tri–Point Triangulation Model

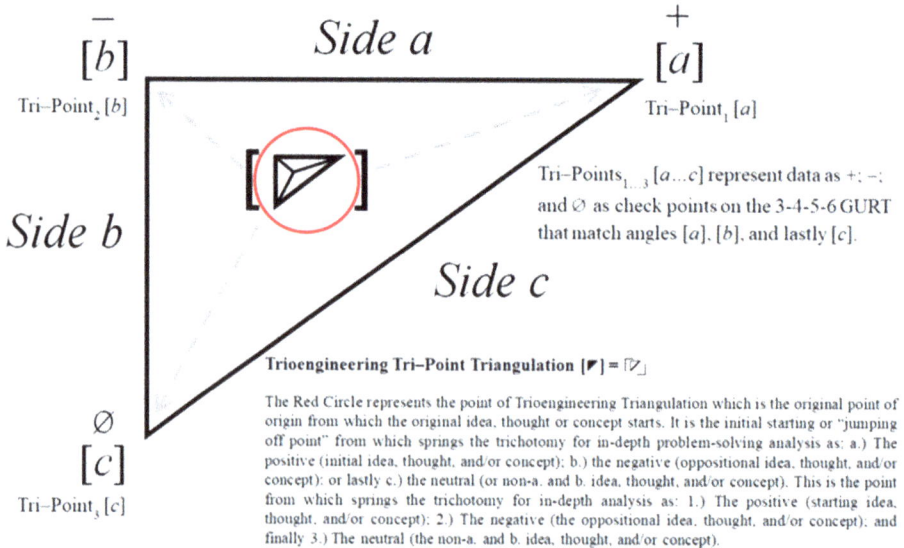

$\overline{[b]}$ **Side a** $\overset{+}{[a]}$

Tri–Point$_2$ [b]   Tri–Point$_1$ [a]

Side b

Tri–Points$_{1...3}$ [a...c] represent data as +; −; and ∅ as check points on the 3-4-5-6 GURT that match angles [a], [b], and lastly [c].

Side c

∅

$[c]$

Tri–Point$_3$ [c]

Trioengineering Tri–Point Triangulation [▶] = [◹]

The Red Circle represents the point of Trioengineering Triangulation which is the original point of origin from which the original idea, thought or concept starts. It is the initial starting or "jumping off point" from which springs the trichotomy for in-depth problem-solving analysis as: a.) The positive (initial idea, thought, and/or concept); b.) the negative (oppositional idea, thought, and/or concept): or lastly c.) the neutral (or non-a. and b. idea, thought, and/or concept). This is the point from which springs the trichotomy for in-depth analysis as: 1.) The positive (starting idea, thought, and/or concept); 2.) The negative (the oppositional idea, thought, and/or concept); and finally 3.) The neutral (the non-a. and b. idea, thought, and/or concept).

Explaining the Internal Structure of the Trioengineering Tri–Point Triangulation Model

The internal structure of the Trioengineering Tri–Point Triangulation Model contains a small Tripositive Tri–Point Triangulation Model in its center. This Model defines the Trioengineering Tri–Point Triangulation Model in terms of "Trichotomous Triangulation Data Points".

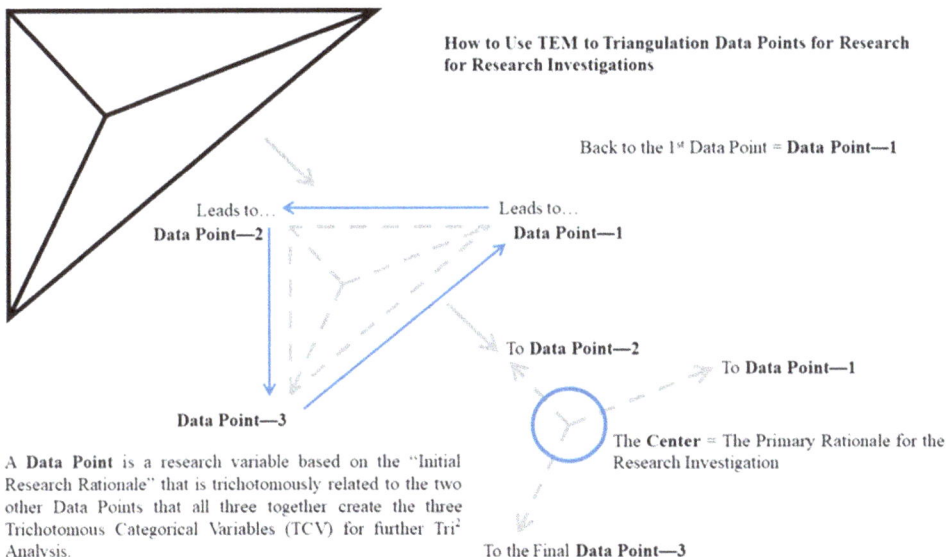

How to Use TEM to Triangulation Data Points for Research for Research Investigations

Back to the 1ˢᵗ Data Point = **Data Point—1**

Leads to… **Data Point—2**

Leads to… **Data Point—1**

To **Data Point—2**

To **Data Point—1**

Data Point—3

The **Center** = The Primary Rationale for the Research Investigation

A **Data Point** is a research variable based on the "Initial Research Rationale" that is trichotomously related to the two other Data Points that all three together create the three Trichotomous Categorical Variables (TCV) for further Tri² Analysis.

To the Final **Data Point—3**

The Internal Characteristics of the Trioengineering Tri–Point Triangulation Model

The internal characteristics of the Trioengineering Tri–Point Triangulation Model are illustrated in the following series of models:

The Trioengineering Tri–Point Triangulation [▼] = ▷⌋
Trichotomous Framework

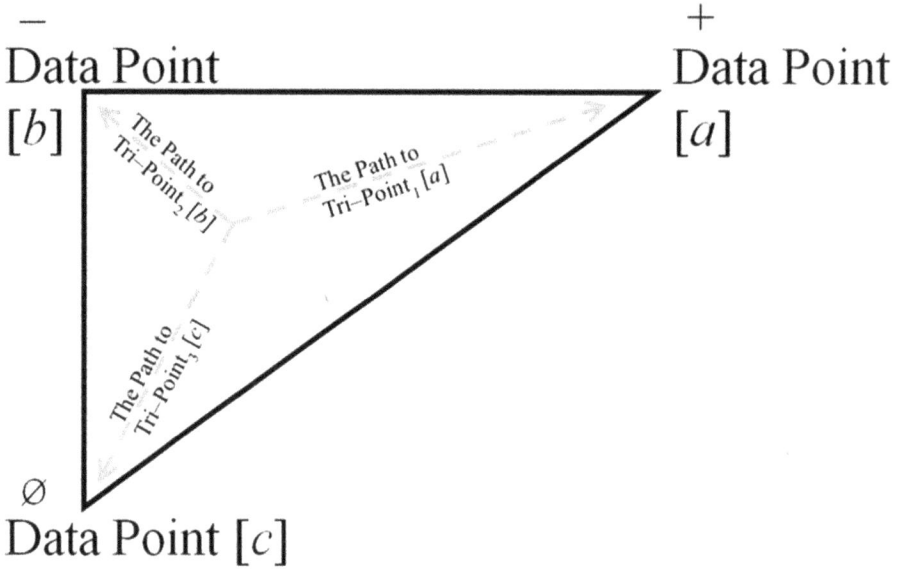

—
Data Point
[b]

The Path to
Tri–Point₂ [b]

The Path to
Tri–Point₁ [a]

+
Data Point
[a]

The Path to
Tri–Point₃ [c]

Ø
Data Point [c]

Triangulation Equation Modeling [TEM] for Triangulated Data In-Depth Analysis

The Model for Triangulated [TEM] is as follows:

Using TEM for Triangulated Data In-Depth Analysis

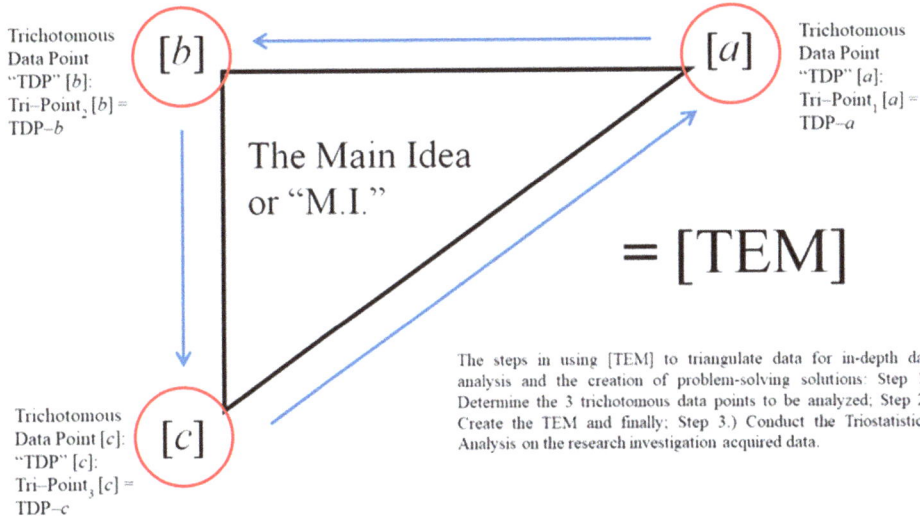

Trichotomous Data Point "TDP" [b]: Tri–Point$_2$ [b] = TDP–b

[b]

[a]

Trichotomous Data Point "TDP" [a]: Tri–Point$_1$ [a] = TDP–a

The Main Idea or "M.I."

= [TEM]

The steps in using [TEM] to triangulate data for in-depth data analysis and the creation of problem-solving solutions: Step 1.) Determine the 3 trichotomous data points to be analyzed; Step 2.) Create the TEM and finally; Step 3.) Conduct the Triostatistical Analysis on the research investigation acquired data.

Trichotomous Data Point [c]: "TDP" [c]: Tri–Point$_3$ [c] = TDP–c

[c]

The Trioengineering Trichotomous Triangulation of Technology

The Trichotomous Triune of E-Learning via Trioengineering [TEM] 4A Metric Algorithm Standards ©

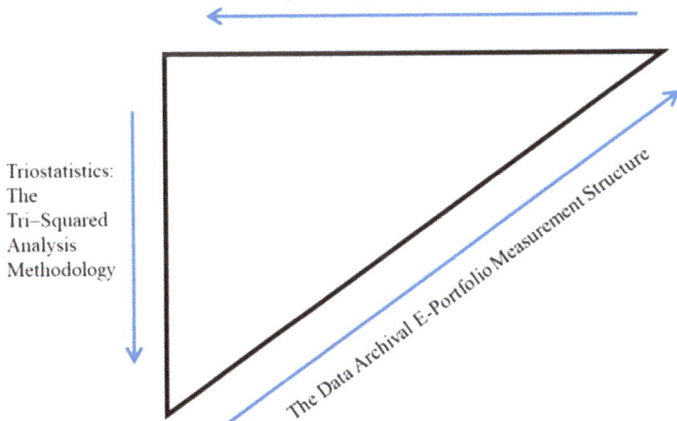

Triostatistics: The Tri–Squared Analysis Methodology

The Data Archival E-Portfolio Measurement Structure

TRIOENGINEERING ™ © *The Problem-Solving Triological Science: The In-Depth Trichotomous Science of the Dynamic 3-4-5-6 Golden Upright Right Triangle for Innovative Problem-Solving.* Osler Studios Incorporated ©, © Copyright 2022 All Rights Reserved.

A further example using the 3-4-5-6 GURT is illustrated in the next series of models.

Trioengineering Triangulation of Technology Using the 3-4-5-6 GURT as an Exemplary Model of Trioengineering for E-Learning

Technology Intensive and Immersive Courses that are actively measurable via the 4A Metric Algorithm Standards ©

The In-Depth measurement Models provided by Triostatistics as evidenced by Triostatistical the following Triostatistical measures: The Tri–Squared Test and [TEM] = Triangular Equation Modeling

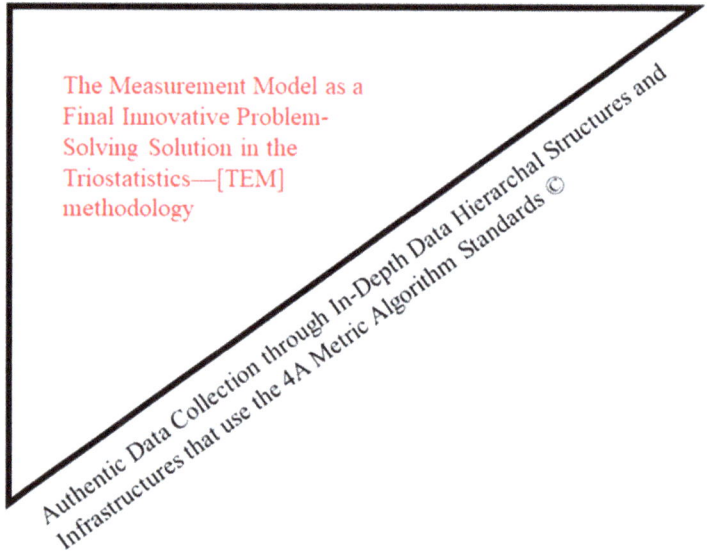

The Measurement Model as a Final Innovative Problem-Solving Solution in the Triostatistics—[TEM] methodology

Authentic Data Collection through In-Depth Data Hierarchal Structures and Infrastructures that use the 4A Metric Algorithm Standards ©

The Trioengineering Instructional Design ADDIE Model Using the 3-4-5-6 GURT as an Exemplary Model of Trioengineering in Deference to its Origins from the Visualus Isometric Cuboid

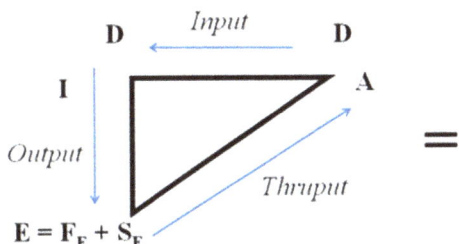

The Visualus Isometric Cuboid ©

The Instructional Design ADDIE Model—
Analysis [+] Design [+] Develop [+] Implement [+] Evaluate = A [+] D [+] D [+] I [+] E = A [+] D [+] D [+] I [+] FE [+] SE
(Note: FE [Formative Evaluation] [+] SE [Summative Evaluation])

The Mathematical Equations Associated with and Related to the Visualus Isometric Cuboid—
The Visualus Volumetric Equation: $v = 9abc$;
The Visualus Isometric Cuboid Surface Area Equation:
$2[9ab + 3ac + 3bc]$ = A [+] D [+] D [+] I [+] E = A [+] D [+] D [+] I [+] FE [+] SE

The Trioengineering 3-4-5-6 GURT in Terms of the Instructional Systems Design (ISD): ADDIE Model to Define to its Unique Terminology

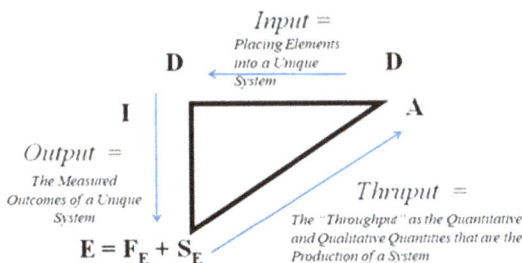

Input =
Placing Elements into a Unique System

Output =
The Measured Outcomes of a Unique System

$E = F_E + S_E$

Thruput =
The "Throughput" as the Quantitative and Qualitative Quantities that are the Production of a System

<u>Transformation of the 3-4-5-6 GURT from the Visualus Isometric Cuboid via Mathematical Definitions</u>

Volume of the Visualus Isometric Cuboid:

$$\frac{1}{2} \cdot \quad = \quad \frac{\frac{1}{2}[L \cdot W \cdot H]}{H} \quad = \quad \frac{\frac{1}{2}[L \cdot W \cdot H]}{H} \quad = \quad \frac{\frac{1}{2}[L \cdot W \cdot H]}{H} \quad = \quad \frac{[L \cdot W]}{2}$$

Area
Calculations:
$[L \cdot W \cdot H]$ = $[B \cdot W \cdot H]$

$\frac{1}{2}[L \cdot W \cdot H]$ = $\frac{1}{2}[B \cdot W \cdot H]$

The Meaning of Trioengineering of
Technology via the Field of E-Learning

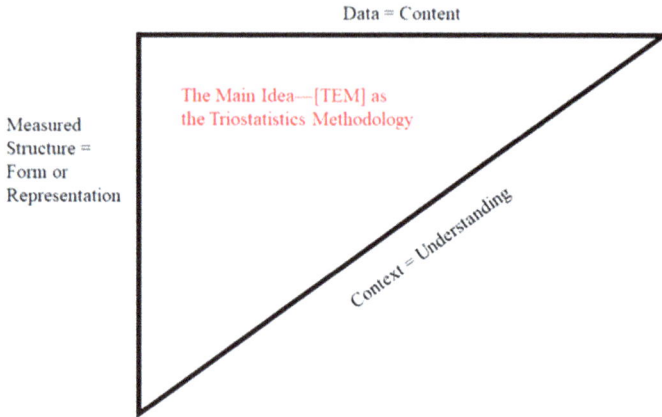

Data = Content

The Main Idea—[TEM] as
the Triostatistics Methodology

Measured
Structure =
Form or
Representation

Context = Understanding

The Base Model of Trioengineering from Research

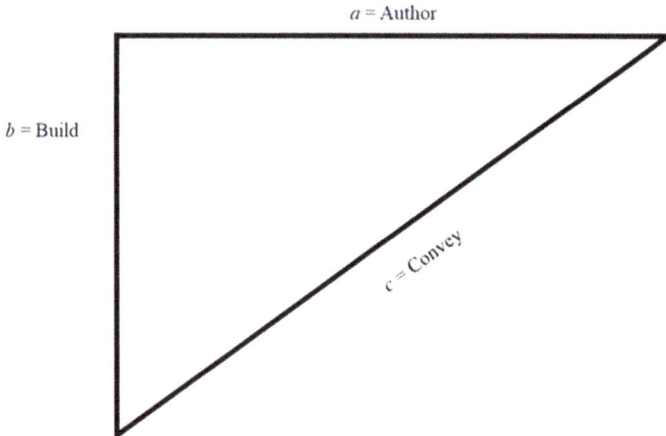

a = Author

b = Build

c = Convey

The Field of "Trioengineering" in terms of Scientific Research: The Trichotomous Research Upright Right Triangle Relevance Index Metric ∇[RIM]

Where "Research Relevance" in terms of a Trichotomous Research Investigation is defined as, "The quality and state of the data as outcomes of an authentic research investigation being – (1.) "Closely Connected" to the investigator's primary assumption and/or posited conjecture; (2.) Appropriately derived with the subject sample under investigation; and (3.) Contemporary based on time, place, and function of the stated research parameters." The final relevance outcome has a trichotomous summative scalar metric of:

The "Tripositive" = "[+]" or "Positive" = "Relevant";
The "Trinegative" = [–] or "Negative" = "Non-Relevant"; or
The "Trineutral" = [0] or "Neutral" as "Zero" = "Non-Existence" (equating to "no value" that is neither "Relevant" or "Non-Relevant").

Application of Trioengineering in Science as the "Complete Tripositive of Trionanotechnology"

Trioengineering in science is emphasized through in-depth analysis and innovative problem-solving. For example, environmental technology-based solutions such as the use of state-of-the-art nanobiotechnology illustrate the use of the science of trichotomy in dynamic and helpful ways under the novel term of "Trionanobiotechnology". Within Trionanobiotechnology there is the observable connection between Trioengineering and its thorough and exhaustive "Tripositive Model". This unique connection can be observed in the following model:

The Model of the Complete Tripositive of Trionanobiotechnology

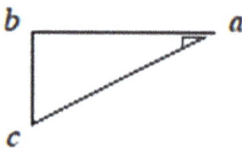

The Trioengineering Model (As a single vector that changes direction to return to its source) a = author; b = build; and lastly to c = convey. That has the following lengths: a to b = 4; b to c = 5; c back to a = 5; or a 3:4:5:6 ratio for the Right Triangle (with "6" indicating the Area).	**The Tripositive Model** (As the three positives that are exactly the same in an "inverted upright right geometric shape" actually a properly represented "Right Triangle" with its respective x and y Cartesian Coordinates at the upper left side.) As illustrated as:

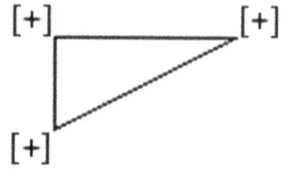

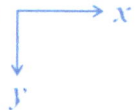

Chapter Six follows and provides examples of how to create solutions using Trioengineering.

TRIOENGINEERING ™ © *The Problem-Solving Triological Science: The In-Depth Trichotomous Science of the Dynamic 3-4-5-6 Golden Upright Right Triangle for Innovative Problem-Solving.* Osler Studios Incorporated ©, © Copyright 2022 All Rights Reserved.

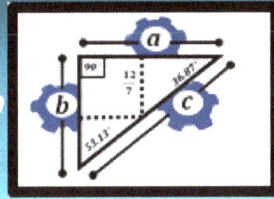

A man shall be satisfied with good by the fruit of his mouth: and the recompense of a man's hands shall be rendered unto him.

Proverbs 12: 14

Trioengineering:
The Trichotomous Science of Innovative Problem-Solving

The trichotomous science term "Trioengineering" is defined as: "The processes, procedures, and products that are the outcomes of problem-solving via "the Mathematical Law of Trichotomy" that yields measurable, scalable, and/or tangible results."

A Symbolic Image for Trioengineering

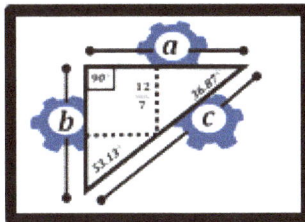

Trithmethic (Pronounced: "Tri"·"rith"·"met"·"tic") Methodology for Trioengineering Using Triology's Trilogic (as "Trilogistics")

The analysis of triangulated data via the "Mathematical Law of Trichotomy" in the context of Trioengineering problem-solving in regards to in-depth data analysis comes in the form of "Tri–Factor" (or "Trifactor") Reflective Analysis". "Tri–Factor Reflective Analysis" is an examination of the connectivity between the Triostatistics Tri–Squared Test (Tri^2) TCVs ("Trichotomous Categorical Variables") and TOVs ("Trichotomous Outcome Variables"). Where each TCV is paired with each TOV across rows and is multiplied using 17th Century mathematician John Napier's calculation methodology termed "Rabdology". Rabdology is calculated in the Standard Tri^2 3 by 3 (or "3 × 3") Table. "Tri–Factor Reflective Analysis" is a Triostatistic in the vein of analytics associated with the Tri–Squared Test that uses a diagonal arithmetic. Tri–Factor Reflective Analysis is a measure of trichotomous commonality as a "reflective connectivity between TCVs and TOVs". The methodology used to calculate Tri–Factor Reflective Analysis is presented in the following "Tri–Factor Reflective Analysis Table" that uses the significant Tri–Squared Test Data

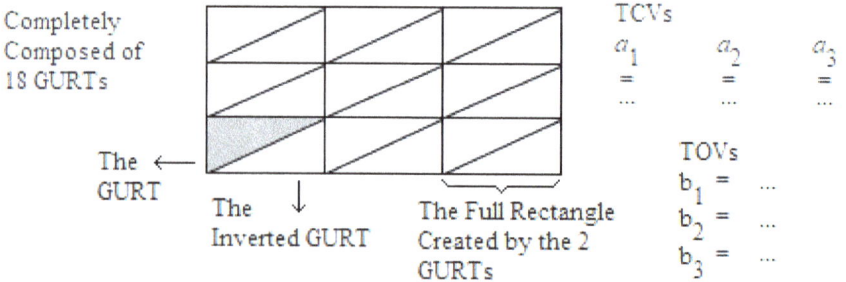

How to Conduct the Tri–Factor Reflective Analysis Operation

The Tri–Factor Reflective Analysis Table sample calculation using sample data for the Tri–Factor Reflective Analysis: "Initial Connective Strength Scale":

TCVs⇨ 1 2 3 ⇩TOVs

	1		2		3		
×	0	1	0	2	0	3	1
×	0	2	0	4	0	6	2
×	0	3	0	6	0	9	3

Total Column Calculations⇨ $1 \times 2 \times 3$ $2 \times 4 \times 6$ $3 \times 6 \times 9$ Grand Total Column Calculation =

$= 6$ $= 48$ $= 162$ $6 \times 48 \times 162$

$= 46{,}656$

Final Results yielded the following – the final Tri–Factor Reflective Analysis "Initial Connective Strength Scale had TCVs had a Tri–Factor Reflective Analysis "Initial Connective Strength Scale" = 1, 2, and 3. The TOVs had a Tri–Factor Reflective Analysis "Initial Connective Strength Scale" = 1, 2, and 3 as well. The final Tri–Factor Reflective Analysis calculation yielded the following final result as **46,656** = Medium, with a on the Reflective Scale that is accurately measured at a **Medium[Low]** with a Tri–Factor Reflective Analysis Reflective Range of: **36,964 to 55,445**.

The Tri–Factor Reflective Analysis Scales

The Tri–Factor Reflective Analysis Scales are as presented below in the section that follows.

The Tri–Factor Reflective Analysis "Initial Connective Strength Scale"

1 = Low;
2 = Medium; and lastly
3 = High.

The Ranges of the Tri–Factor Reflective Analysis—"Ranged Reflective Scale"

[0 to 36,963] = <u>**Low**</u>, with the in-depth ranges that are measured accurately according to the following within *"Low Reflective Scalar Range"*—(**Low[Low]** with a Tri–Factor Reflective Analysis Reflective Range of **0 to 9,240**), (**Low[Medium]** with a Tri–Factor Reflective Analysis Reflective Range of **9,241 to 18,481**), and (**Low[High]** with a Tri–Factor Reflective Analysis Reflective Range of **18,482 to 36,963**); followed by—

[36,964 to 73,926] = <u>**Medium**</u>, with the in-depth ranges that are measured accurately according to the following within *"Medium Reflective Scalar Range"*—(**Medium[Low]** with a Tri–Factor Reflective Analysis Reflective Range of **36,964 to 55,445**), (**Medium[Medium]** with a Tri–Factor Reflective Analysis Reflective Range of **55,446 to 64,685**), and (**Medium[High]** with a Tri–Factor Reflective Analysis Reflective Range of **64,686 to 73,926**); and lastly—

[73,927 to 110,889+] = <u>**High**</u>, with the in-depth ranges that are measured accurately according to the following within *"High Reflective Scalar Range"*—(**High[Low]** with a Tri–Factor Reflective Analysis Reflective Range of **73,927 to 83,167**), (**High[Medium]** with a Tri–Factor Reflective Analysis Reflective Range of **83,168 to 97,028**), and (**High[High]** with a Tri–Factor Reflective Analysis Reflective Range of **97,029 to 110,889+**).

What Tri–Factor Reflective Analysis is Actually Design to Measure

Tri–Factor Reflective Analysis is designed to measure the efficacy of Triostatistical analysis and analytics as an advanced post hoc methodology particularly aimed at in-depth trichotomous metrics that directly aggregate data from participant-based metrics such as the traditional Tri–Squared Test. Tri–Factor (or "Trifactor") Reflective Analysis is a post hoc progressive and forward-thinking analysis. "Tri–Factor" is a trichotomous analytic that determines the unique connective strength and level of reflective range between Tri–Squared Test trichotomous categorical and outcome variables as "dynamic trichotomous research investigation factors".

Trioengineering in Terms of Learning

Trichotomous Trioengineering in terms of learning is designed to aid the trichotomous scientist to think in terms of "How to Think in Threes" (especially in regards to "Triangular Equation Modeling" = [TEM]). "Thinking in Threes" consists of: The initial thought, idea, and/or concept; The oppositional thought, idea, and/or concept; and neither the initial or oppositional thought, idea, and/or concept. Trioengineering also aids in the following areas of Triology (the Triological Science):

1.) *"The Science of Decision-Making"*;
2.) *"The Science of Ascertaining Experiences"*; and
3.) *"The Science of Analyzing and Evaluating Perspective"*.

In terms of "Trioengineering in Learning" the following areas have had great impact from the discipline:

1.) *Psychometric Trioengineering for Virtual Learning Measurement*;
2.) *Interactive Triostatistics for Virtual Learning Measurement*;

3.) *Advanced Triostatistics for Virtual Learning Measurement*;
4.) *E-Learning Standardization and Measurement*: **and**
5.) *Instructional Leadership for Virtual Learning Leadership*.

Trioengineering in terms of learning provides dynamic organizational concept mapping falls under the novel term of "**Triflow ©**" (in the vein of "Trichotomous Thinking" in regards to "Triangular Equation Modeling" = [TEM]). "Triflow" is a portmanteau of the two terms "Trichotomous" and "Workflow". It is defined as "Trichotomous Thinking" that is smooth from a holistic perspective, that is centered around a central "Main Idea" that is a solution as an initial thought, idea, and/or concept. Triflow is trichotomously divided into three main sectors: "parameter or sector a" opposed by "" (in an article published by the author—Osler, 2019)".

"Trioengineering Learning" is active in data analysis modeling of more in-depth Triostatistical analysis in the form of "***Trichotomous Data Triangulation*** (or "**TDT**") in respect to "Triangular Equation Modeling = [TEM]). TDT is referred to in the abbreviated form of "**Trichotomous Triangulation**" or "**Trioengineering Triangulation**" or "**Trioengineering Tri–Point Triangulation**" under the full title of "Trichotomous Triangulation: Trioscientific Advanced Triostatistical Past Hoc Analysis of Significant Tri–Squared Input and Output Based on Initial Research Variables". TDT as has the following models (as illustrated in the previous chapter) illustrated as equal (side by side) in the next section for more clarity.

TDT Tri-Point Graphical Models: The Trioengineering Tri–Point Triangulation Model = Internal Structure of the Trioengineering Tri–Point Triangulation Model = The Internal Characteristics of the Trioengineering Tri–Point Triangulation Model

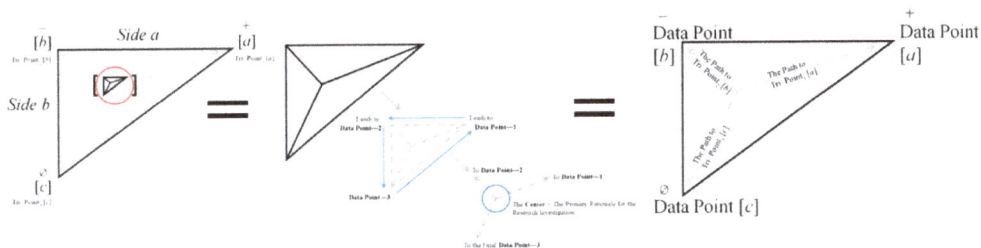

Further mathematical and graphical models can be illustrated to further define the unique relation ships that exist between GURT angles and the primary utilization in the development of innovative Trioengineering solutions. Some Trioengineering solutions can become the basis of trichotomous inquiry. This is illustrated in the final trichotomous inquiry model in the form of questions 1 through 3 in the section that follows.

The Graphical and Mathematical Models for TDT, Trichotomous Tri–Point Triangulation via the Triangulation Equation Model [TEM]

$$[a + b + c] =$$

$$[(angle)a + (angle)b + (angle)c] =$$

$$[\llcorner a + \llcorner b + \llcorner c] =$$

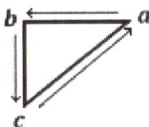

Where, in the model above:

---

a = Factor a;
b = Factor b; and
c = Factor c.

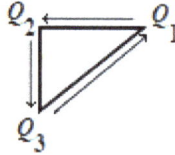

Where, in the model above:

Q_1 = Question 1;
Q_2 = Question 2; and
Q_3 = Question 3.

The Trioengineering Relationship between the Visualus Isometric Cuboid and the 3-4-5-6 GURT as the "Total Trichotomous Triangulum" or "3T" of the Isometric Cuboid

The Trioengineering "Total Trichotomous Triangulum" (or "3T") is a descriptive set of terms for the 3-4-5-6 GURT that is directly derived from the Visualus Isometric Cuboid. The GURT comes from the front face of the Isometric Cuboid when it is divided in half diagonally. Additionally, the Standard 3 by 3 Table of the Triostatistics Tri–Squared Test is derived is the same manner as it is the total front face of the Isometric Cuboid applied to trichotomous statistical measurement and data analysis. The graphical model that follows illustrates the unique relationship between the Isometric Cuboid, the GURT, and the Trigmoid symbol, function, and equations.

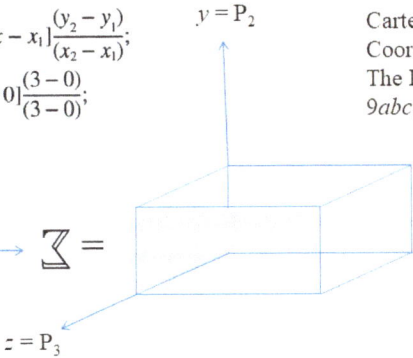

$$\triangledown[\zeta] = \nabla y = y_1 + [\nabla x - x_1]\frac{(y_2 - y_1)}{(x_2 - x_1)};$$

$$\triangledown[\zeta] = \nabla y = 0 + [3 - 0]\frac{(3 - 0)}{(3 - 0)};$$

$$\triangledown[\zeta] = \nabla y = 3$$

Cartesian Coordinates "xyz" = $[P_1 \cdot P_2 \cdot P_3]$ = The Isometric Cuboid Volume = v = $9abc$, where, $a = x$; $b = y$; and $c = z$

The Trigmoid = 3-4-5-6 GURT $\longrightarrow$ ζ =

$y = P_2$

$x = P_1$

$z = P_3$

= $[P_1 \cdot P_2 \cdot P_3]$ = The Visualus Isometric Cuboid Volumetric Areas

Chapter Seven follows and describes the origins of Trioengineering.

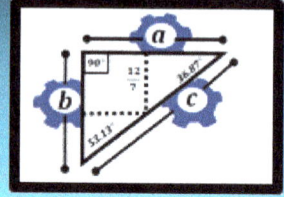

For the vision is yet for an appointed time, but at the end it shall speak, and not lie: though it tarry, wait for it; because it will surely come, it will not tarry.

Habakkuk 2: 3

Technology Engineering: A Mathematical Model of Interactive Interelectronic Ideation Courseware Design

A mathematical model that illustrates the application of Technology Engineering is the construction of ergonomic courseware using the **Ergonomic Multimedia Courseware Equation** © (or "*emc*") (Osler, 1996). The equation is the outcome of research concerning Interactive Technology designed to enhance learning. It is written:

$$\int^{emc} = [f^{sb}(b) + (C^s + H^u + N^{ag} + L^{bt} + L^{bg})^\Delta] + S^{mt}$$

The purpose of the mathematical *emc* equation is to provide an illustration of how ergonomically–designed courseware is structured; and provide an example of how each ergonomic factor is applied and utilized in conjunction with the other ergonomic factors in the overall courseware design. When the ergonomic factors in the equation are properly combined, they create an interactive user interface that has the following characteristics: ease of use, perpetuation of interest, delivery of content knowledge, active engagement, and a high locus of control.

Metacognetic Mechanics, Technology Engineering, and Optimal Instruction

As stated in Chapter 2 the bridging the terms "Meta" and "Cognetic" creates the new term "Metacognetic". **"Metacognetics"** are defined as "The discipline concerned with the creation of instructional delivery methods that go beyond traditional instructional strategies and incorporate innovative learning strategies such as: ergonomics, multiple intelligences, best instructional practices, and metagraphics to improve the retention and transfer of knowledge in the learning environment". **Metacognetic Mechanics** are defined as, "Instructional methodologies that fuse ergonomic principles with purposeful design to engineer a union of content and method of delivery designed to increase the retention and transfer of content knowledge". **"Mechanics"** is the branch of physics concerned with the behavior of physical bodies. The discipline of Physics originated with Mechanics. Physics is grounded in Mechanics which provides an extensive body of knowledge about the natural world. The study of Mechanics is also a central part of **Technology** (the practical application and use of knowledge towards a particular task to produce crafts, systems, and tools that allow humanity to better interact with their environment). Through Technology, the discipline of Mechanics is often known as either "Engineering" or "Applied Mechanics". In the discipline Engineering, Mechanics is used to design and analyze the behavior of machines, mechanisms, and structures. When Mechanics are used as "Applied Mechanics" a discipline is created that is referred to as "Theoretical and Applied Mechanics". Theoretical and Applied Mechanics is a branch of the physical sciences. In the "Theoretical Sciences", the discipline of Theoretical and Applied Mechanics is useful in the scientific formulation of new ideas and theories, the discovery and interpretation of phenomena, and the development of experimental and computational tools.

In "Metacognetic Mechanics", the term "Mechanics" is used in a similar fashion to the discipline of Theoretical and Applied Mechanics. The term "Mechanics" in "Metacognetic Mechanics" literally means to formulate new ideas and theories to explain the physical development and structure of Metacognetic concepts and principles. This is exemplified in the practice of "Technology Engineering" an active branch of Metacognetic Mechanics. Metacognetic Mechanics is also concerned with the thought processes that directly apply to developing instructional methods and strategies that enhance traditional instruction. Metacognetic Mechanics specific "metacognitive" ("thinking about the process of thinking") processes regarding the development of specialized digital systems to enhance the retention and transfer of knowledge" is the "Technology Engineering" methodology. Technology Engineering develops specific principles that expound the examination, design, and construction of instructional tools that are based upon the metacognetic engineering principles that yield interactive Metametric systems designed to augment, enhance, and increase learning. The discipline of Metacognetic Mechanics is structured in the following manner: Optimal Instruction, defined as the branch of Metacognetic Mechanics which studies the optimally–effective instructional strategies that incorporate the functions, similarities, and collaboration of the specific instructional terms, definitions, and learning strategies. The instructional strategies covered in Optimal Instruction include: Cognitive Economy, Cognitive Loading, Retention, Transfer, Human Computer Interaction, Content Authoring, Interactive Learning, Metametric, and Learning Tools. Metacognetic Mechanics uses mathematical modeling to explain its principles, concepts, and methods. Technology Engineering exemplifies the mathematical model of Metacognetic Mechanics as it yields a tangible solution that is an Interactive Technology learning tool Note: Technology Engineering is the instructional delivery methodology and Optimal Instruction is the instructional engagement strategy in which the instructional delivery methodology is put to use.

Metacognetic Mechanics is the overall discipline that actively uses the both Technology Engineering and Optimal Instruction to create a more effective instructional environment. Through Visualus, an equation can be calculated that illustrates the relationship between Metacognetic Mechanics, Technology Engineering and Optimal Instruction. This equation is called "**The Metacognetic Mechanics Equation ©**". It is written as follows:

$$\mathbf{MM = TE + OI}$$

The Visualus Metacognetic Mechanics Equation is defined by the following Visualus Solutions (the step–by–step methodology used to calculate these and other types of Visioneering Volumetric Solutions are presented in Chapter 12):

$$\left[\coprod_{M}{}^{M}{}_{eme} \equiv \coprod^{MM_{eme}} \equiv \coprod_{M}{}^{M}{}_{eme} \right] =$$

$$\left[\coprod_{E}{}^{T}{}_{emc} \equiv \coprod^{TE_{emc}} \equiv \coprod_{E}{}^{T}{}_{emc} \right] + \left[\coprod_{I}{}^{O}{}_{smc} \equiv \coprod^{OI_{smc}} \equiv \coprod_{I}{}^{O}{}_{smc} \right]$$

Where,

MM = *Metacognetic Mechanics,*   [*eme* = Education Metametric Equation]
TE = *Technology Engineering,*   [*emc* = Ergonomic Multimedia Courseware]
OI = *Optimal Instruction,*   [*smc* = Supportive Course Materials]

In the next section, Technology Engineering is explored and examined in detail via the in–depth research that was conducted on explicative mathematical models used to clarify the various concepts and operating principles that are unique to Technology Engineering as an Interactive Technology.

***

The Mathematical Model Illustrating the Summation of the Product of Technology Engineering in Terms of the Construction of the Integrated Ergonomic Multimedia Courseware Interface:

$$\sum_{emc} = \int^{emc} \left[f^{sb}(b) + G^{ui}_{\Delta} \right] + I^{t}$$

This equation is defined as: **The Sum of Ergonomic Multimedia Courseware ©** is equal to the **Integrated Ergonomic Courseware Interface Equation ©** (also called the "**Ergonomic Interface Equation ©**" abbreviated as "**EIE**"). The EIE is designed to address the manner in which effective Ergonomic Multimedia Courseware is structured. The next series of mathematical formulae further illustrate interactive ergonomic courseware architecture. The complete design of a Graphic User Interface constructed to maximize content delivery is evident in the **Interface Change Operation ©** $\left[G^{ui}{}^{\Delta}_{\nabla} \right]$ indicating the changes in interface design using delta [Δ] to indicate overall change and nabla [∇] to indicate gradual change. The Interface Change Relation is equal to the Integrated Ergonomic Courseware Interface Equation as illustrated in the following equation:

$$G^{ui}{}^{\Delta}_{\nabla} = \int^{emc} \left[f^{sb}(b) + G^{ui}_{\Delta} \right] + I^{t}$$

The use of the Transitive Property of Equality then indicates:

$$\int^{emc} \left[f^{sb}(b) + G^{ui}_{\Delta} \right] + I^{t} = \sum_{emc}$$

***

TRIOENGINEERING ™ © *The Problem-Solving Triological Science: The In-Depth Trichotomous Science of the Dynamic 3-4-5-6 Golden Upright Right Triangle for Innovative Problem-Solving.* Osler Studios Incorporated ©, © Copyright 2022 All Rights Reserved.

Thus, the Sum of Ergonomic Multimedia Courseware (where the index [*i*] is the Ergonomic Multimedia Courseware Equation $\left(\int^{emc} = emc\right)$, thus, $\left(\sum_{i\,=\,emc} = \sum_{emc}\right)$ is equal to the final design of the Graphic User Interface. Thus, $\sum_{emc}$ is the Interface Change Operation $\left[\,G^{ui}{}_{\nabla}^{\triangle}\,\right]$. This relationship is mathematically written in the following manner:

$$\sum_{emc} = G^{ui}{}_{\nabla}^{\triangle}$$

The equation leads to a more carefully observation the entire process involved in the **emc** mathematical model. The **Product of Technology Engineering** © is the product or outcome of the multiplied summation of the Isometric Cuboid Planar Vectors *a*, *b*, and *c* (exemplified in the Essential Equation of Visualus: **v = 9abc**). This clearly illustrates the Innovative Problem–Solving Model of Inventive Instructional Design as a dynamic interactive instructional solution to some specified problem in the learning environment.

Summarizing the Mathematical Definition, Origin, and Purpose of Technology Engineering

The Technology Engineering methodology integrates and seamlessly infuses interactive technology with content and curriculum. It was developed to provide an empowering and effective instructional strategy that makes use of the various modes of instruction available through Human Computer Interaction. Technology Engineering is the combination of two interrelated concepts combined together as a whole. "Technology" is defined in this case, as the use of courseware and computers as a means of delivering instruction. "Engineering" is likewise defined as the problem–solving application of technology as a means of delivering information. The two concepts unified together create the title "Technology Engineering" which is the product of the two terms collaboratively combined to describe the methodology that includes the application of interactive technology to the learning environment.

The relationship between Technology and Engineering can be clearly illustrated by the mathematical equation,

$$TE_I = \prod$$

Where, Technology Engineering (TE_I) is represented as the Product ($\prod$) of the combined sets of "Technology" (T) and "Engineering" (with the symbol "E_I" representing "The Engineering of Information" or "Information Engineering" (which is the careful and specific process of constructing, designing, and disseminating knowledge to enhance learning). The Product of Technology Engineering is Information Engineered as content for curriculum structured to take advantage of applied Interactive Technology that uses the practice of "Impacting Technology" in the learning environment. The goal is to create an "Interactive Cognitive Economy" in the learning environment that empowers learners and facilitates the retention of subject matter and the transfer of specific content knowledge.

The Product of Technology Engineering is comprehensively and mathematically represented by calculating the **Technology Engineering Solution** © (covered in detail as a Visualus Calculation in Chapter 12). The relationship between the Product of Technology Engineering and the Technology Engineering Solution is one of equality. The Technology Engineering Solution is the mathematical process used to show the unity of Technology with Engineering that produces as a definitive outcome the Ergonomic Multimedia Courseware Equation (which is actually the Product of Technology Engineering). The Product of Technology Engineering mathematical equation has the following definitive categorical areas:

$$TE_I = \prod$$

1.) **"Technology"** = symbolized as (*T*) in the equation for the "Metametric Instructional **T**ool", with the term "Technology" being broadly defined in this equation as the use of courseware and computers as a means of delivering instruction.

2.) **"Engineering"** = symbolized as (E_I) in the equation for "**I**nformation Engineering", and is the process used to create the Metametric Instructional Tool, with the term "Engineering" being broadly defined in this equation as the problem–solving application of technology as a means of delivering information.

Chapter Eight follows and describes advanced ergonomics in terms of Visualus, further mathematically defining Technology Engineering.

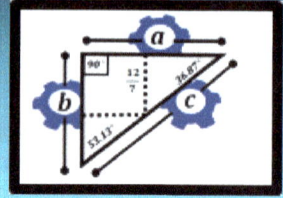

Every good gift and every perfect gift is from above, and cometh down from the Father of lights, with whom is no variableness, neither shadow of turning.

James 1: 17

$$\nabla[\mathfrak{Z}] = \nabla y = y_1 + [\nabla x - x_1]\frac{(y_2 - y_1)}{(x_2 - x_1)};$$

$$\nabla[\mathfrak{Z}] = \nabla y = 0 + [3 - 0]\frac{(3-0)}{(3-0)};$$

$$\nabla[\mathfrak{Z}] = \nabla y = 3$$

The Trigmoid = 3-4-5-6 GURT $\rightarrow \mathfrak{Z} =$

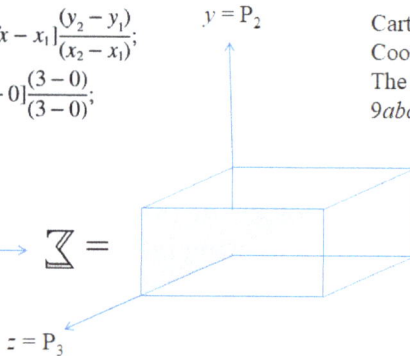

$y = P_2$

Cartesian Coordinates "xyz" = $[P_1 \cdot P_2 \cdot P_3]$ = The Isometric Cuboid Volume = v = $9abc$, where, $a = x$; $b = y$; and $c = z$

$x = P_1$

$z = P_3$

= $[P_1 \cdot P_2 \cdot P_3]$ = The Visualus Isometric Cube Volumetric Are

In the above illustrated graphic $[P_1 \cdot P_2 \cdot P_3]$ = "[Act]" (for "Actuality") is defined as follows: "Actuality" is equal to the concentrated "Triposition" ("The focused concentration of a triple positive trichotomous tricoordinate factor mathematically indicated as '[]' for the purposes of trichotomous mathematics") of $[P_1]$ = "Position" = Cartesian Coordinate "x" representing "Place" ; $[P_2]$ = "Perception" = Cartesian Coordinate "y" representing "Sensation"; and $[P_3]$ = "Perspective" = Cartesian Coordinate "z" representing "Awareness", for the express purpose of representing "Trichotomous Power". Therefore, "Actuality" according to the abovementioned graphical model has the following mathematical definition in deference to the aforementioned Tripositive mathematical definitions: "$[Act] = [P_1 \cdot P_2 \cdot P_3] = [P_{1...3}]^{3}$". As

---

such, "Actuality is equal to the concentrated Position times Perception times Perspective Cubed". In terms of mathematical clarity, the following is now applicable:

1.) The **Trioengineered** *"__Position__"* is now defined in terms of mathematical Triposition as "**Ability**" that is defined as an individual's level of spatial awareness of what around them in terms of physical and tangible special psychomotor insight of what is transpiring around them, and the determination of exactly where they are ("spatially") in terms of exactly where they are (in terms of their "respective spatial existence") at the present time;

2.) The **Trioengineered** *"__Perception__"* is now defined in terms of mathematical Triposition as "**Aptitude**" that is defined as an individual's level of comprehension or understanding of knowledge (cognitive/cognition), skills (psychomotor/tangible application towards), disposition (attitude/opinions and emotions), and interactions (relations/relationships with); and

3.) The **Trioengineered** *"__Perspective__"* is now defined in terms of mathematical Triposition as "**Action**" that is defined as an individual's attitudinal level regarding any particular part or aspect of existence.

The aforesaid can be actively defined by *"Osler's General Contextuality Actuality Maxim"* that is mathematically defined as: "Actuality as '[A*ct*]' = $[P_1 \cdot P_2 \cdot P_3] = [P_{1...3}]^3$".

Trioengineering as Educational Science (as "Eduscience" from the Osler 2013 research article entitled "The Psychological Efficacy of Education as a Science through Personal, Professional, and Contextual Inquiry of the Affective Learning Domain")

Trioengineering as Educational Science or "Trioengineering Eduscience" has following three aspects as the trichotomy of Eduscience Subtopics:

1.) *"Eduscience* **as an Art";**
2.) *"Eduscience* **as a Practice"; and**
3.) *"Eduscience* **as Outcome-Driven Research".**

In the unique discipline of "Trioengineering Eduscience" information is mainly contextual with the understanding that information delivered within the circumstantial confines of learning is primarily, essentially, and fundamentally situational. This then means that each "Learning Environment" in the confines of information delivery as a unique information transfer procedure/process, and information encoding intake (on the part of the "participant"/"learner") is dependent upon the following:

a.)   The uniqueness of the information disseminator—
       (as the teacher/instructor/professor);

b.)   The uniqueness of the receiver—
       (as the student/learner/participant/apprentice/mentee); and

c.)   The uniqueness of the structure—
       (as the learning environment/classroom/situation).

The unique key to all of the aforementioned being the motivation to learn (by the student/learner/participant/apprentice/mentee) and the simplicity in which and how the specific information is delivered (by the teacher/instructor/professor) and works within the confines of the unique learning situation (as the learning environment/classroom/situation).

The Trioengineering Tripositional Analysis Methodology

The Trioengineering Tripositional Analysis Methodology is composed of the following Five Phases—

Phase One: Topic Selection (as an inquiry, solution, concept, and/or as an idea);

Phase Two: The Initial Topic Polarization (Separation into the three trichotomous areas);

Phase Three: Determination of the Trichotomous Interposition Nullification (as the Opposition to the Initial inquiry, solution, concept, and/or as an idea);

Phase Four: Construction of the Final "Triangular Equation Model" or "[TEM]" on the Topic Using the Outcomes of Phases Two and Three respectively; and lastly

Phase Five: Present and Define the Final [TEM].

The equality of the unique relationships inherent to the *"Trioengineering Tripositional Analysis Methodology"* is clearly illustrated in the graphical illustrated model that immediately follows and visibly exhibits the clearly defined unique triple equality that exists between, within, and around the triple trichotomy of the GURT in three uniquely identical GURTs derived out of the exclusive and exceptional characteristics of the 3-4-5-6 Golden Upright Right Triangle.

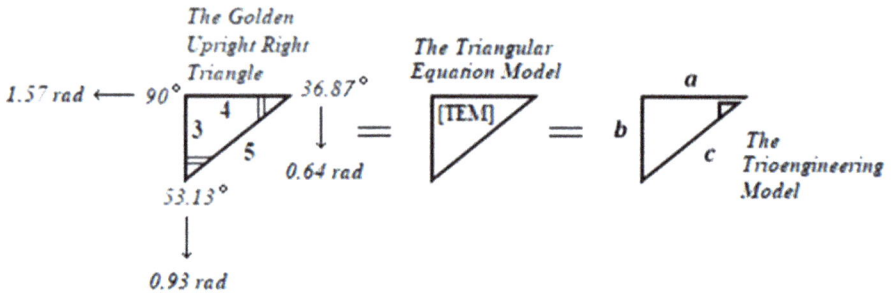

The Golden Upright Right Triangle • The Triangular Equation Model [TEM] • The Trioengineering Model

1.57 rad ← 90°
36.87°
↓ = 0.64 rad
53.13°
↓
0.93 rad

Chapter 9 follows and describes the process of "Trichotomous Research" as it is comprehensively defined via Trioengineering.

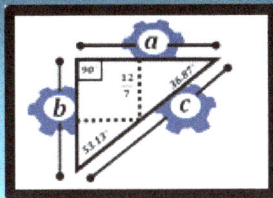

Give instruction to a wise man, and he will be yet wiser: teach a just man, and he will increase in learning.

Proverbs 9: 9

Trioengineering Notation in Action: Trioengineering Notation Defined

∇ or $\nabla N = \displaystyle\sum_{i=1}^{3}$ for $\left.\begin{array}{l} a = 4; \\ b = 3; \\ c = 5; \text{ and} \\ A = 6. \end{array}\right\}$ $= \nabla abc =$ [triangle: build, author, convey] as [TEM] =

[triangle labeled 1, 2, 3]   expressed also as the Tripositive = ∇ = [Tri].

Where,

N = "Name" of the measurement/concept/solution; and ∇ = "The 3-4-5-6 Golden Upright Right Triangle" or "Triune" in action as the "Trioengineering Trichotomate" and " Trichotomous Triangulation" = "Trioengineering Trichotomate".

Thus,

$\nabla N = \displaystyle\sum_{i=1}^{3} [N] \equiv \nabla Name_1 + \nabla Name_2 + \nabla Name_3 = [\ \nabla Name_1;\ \nabla Name_2;$ and $\nabla Name_3]$ for Primary Main Name in:

∇I_1 = "Trichotomous Triangle Inculcation" ≡ , ∇M = Middle of
$\nabla[M] = \nabla[\mathbf{\exists}]$;
∇I_2 = "Trichotomous Triangle Intercalation" ≡ $\nabla[TEM]$; and
∇I_3 = "Trichotomous Triangle Interpolation" ≡ $\nabla[$"Side c"$]$.

In addition, the following applies:

$\nabla...$ = "The Trioengineering Trichotomation of: "..." "; (Note: the term "Trichotomation" is defined as the separation into units of 3 based upon "The Mathematical Law of Trichotomy") and
∇N = "The Trioengineering Trichotomation of: "N" "; with the following mathematical definitions that are applicable to the aforementioned trichotomy—

∇Inculcation$_1$ = " ∇Interior" = Inside of the 3-4-5-6 GURT as the [M.I.] = [Main Idea] = $\nabla[\mathbf{\exists}]$ = The Overall Overriding Subject;
∇Intercalation$_2$ = " ∇Exterior" = "Trioengineering Trichotomation" = ∇ [TEM] ;
∇Interpolation$_3$ = " ∇ Ulterior" = The Trigmoid Function for "Side c" of the 3-4-5-6 GURT = The Final Solution as Side c = "*convey*".

Note the following:

$$\nabla[\mathbf{\exists}] = \nabla y = y_1 + [\nabla x - x_1]\frac{(y_2 - y_1)}{(x_2 - x_1)};$$

$$\nabla[\mathbf{\exists}] = \nabla y = 0 + [3 - 0]\frac{(3 - 0)}{(3 - 0)};$$

$$\nabla[\mathbf{\exists}] = \nabla y = 3$$

$$V = \overset{3}{\underset{i=1}{V}} = \text{[Tri] as "Trioengineering Trichotomous Triangulation"} =$$

"Trichotomation" = The combination of $I_1...I_3...$ This is further defined with the following 4 Primary Cartesian Coordinates which are $\{(x_1 = 0, x_2 = 3); (y_1 = 0, y_2 = 3); (x_1 = 0, y_1 = 0);$ and $(x_2 = 3, y_2 = 3;)$ as $V[3:3]$ with the same Cartesian Coordinates except for the last for $V[4:3]$ that are $(x_3 = 4, y_3 = 3;)\}$ and mathematically in the following manner:

$$\overset{3}{\underset{i=1}{V}}[3] = Vy = y_1 + [Vx - x_1]\frac{(y_2 - y_1)}{(x_2 - x_1)} = V[\text{TEM}] = [\text{TEM}], \quad \text{for}$$

$$= [\, V\text{Part 1} = VP_1; \; V\text{Part 2} = VP_2; \text{ and } V\text{Part 3} = VP_3];$$

$$\overset{3}{\underset{i=1}{V}}[3] = \text{Part 3} = (\text{Main Idea} - \text{Part 1} + \text{Part 2});$$

$$\overset{3}{\underset{i=1}{V}}[3] = \;\;\text{at (0, 0) to (3, 3) with } Vf(x) = Vy = Vy\text{-axis} =$$

... = Ordinate;

$$= Vy = 0 + [3 - 0]\frac{(3-0)}{(3-0)} = 3(3/3) = 3(1) = 3;$$

$\nabla y = 3 = \nabla[\text{TEM}] =$

(0, 3) (3, 3)
(0, 0)

$=$

(0, 3) Part 1 (3, 3)
Main
Part 2 Idea
Part 3
(0, 0)

$=$

∇P_1
b ← a
∇P_2 [M.I.]
∇P_3
c
$\rightarrow \nabla[\text{M. I.}];$

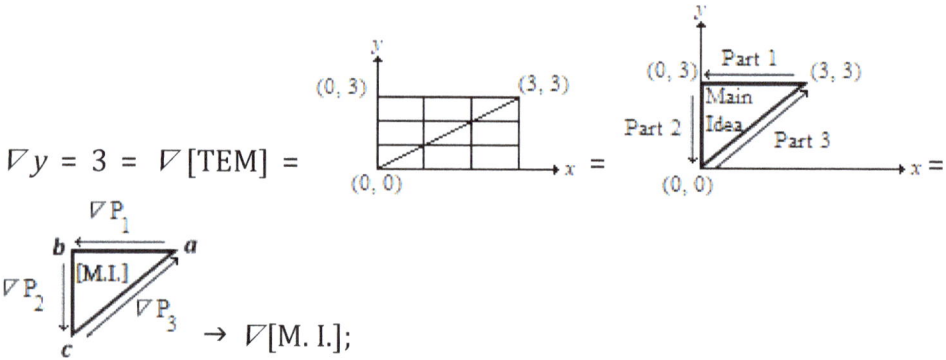

From rectangle to Square: The Coequality of [3:3] with [4:3] is graphically expressed as:

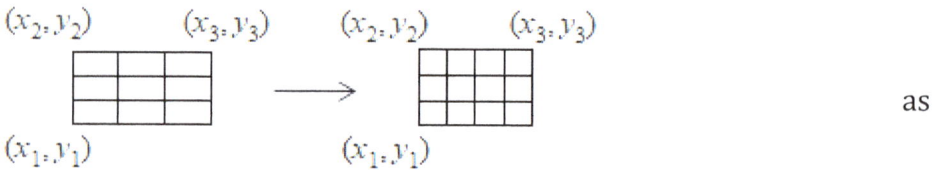

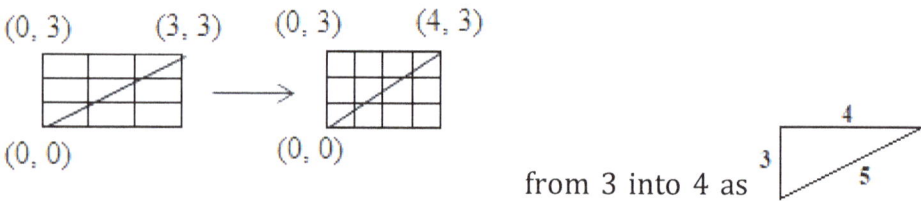

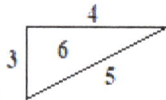

(x_2, y_2) (x_3, y_3) (x_2, y_2) (x_3, y_3)

$\longrightarrow$

(x_1, y_1) (x_1, y_1)

as

(0, 3)   (3, 3)   (0, 3)   (4, 3)

$\longrightarrow$

(0, 0)   (0, 0)

from 3 into 4 as

4
3
5

and A = 6 into

4
3   6
5

. Thus, the following (within the previously specified specifics of "Trioengineering Notation" holds true with a final equality to a "Tripositive Trigmoid", observe the following graphic models that emphasize Trioengineering Notation:

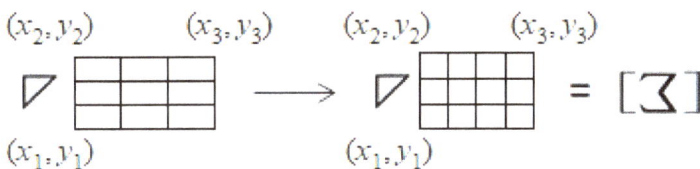

(x_2, y_2) (x_3, y_3) (x_2, y_2) (x_3, y_3)

$\triangleright$ ▦ $\longrightarrow$ $\triangleright$ ▦ $= [\mathbf{3}]$

(x_1, y_1) (x_1, y_1)

into

$(0, 3)$ $(3, 3)$ $(0, 3)$ $(4, 3)$

$\triangleright$ ▦ $\longrightarrow$ $\triangleright$ ▦ $= [\mathbf{3}]$.

$(0, 0)$ $(0, 0)$

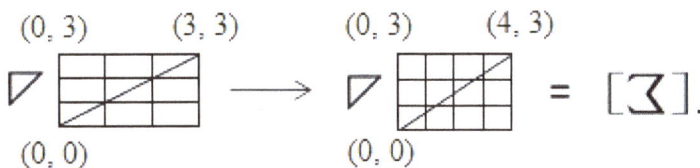

Understanding the Mathematical Law of Trichotomy to Better Grasp the Overall Rationale of Trioengineering

The general definition of the term "Trichotomy" and the "Mathematical Law of Trichotomy" are defined by several researchers in the 2012 Journal on Mathematics research article published by the author as follows:

"The term is pronounced ['trahy-kot-uh-mee'], spelled "trichotomy", and is a noun with the plural written form "trichotomies". A "Trichotomy" in terms of philosophy can be referred to as a threefold method of classification. Philosopher Immanuel Kant adapted the Thomistic acts of intellect in his trichotomy of higher cognition — (a) understanding, (b) judgment, (c) reason — which he correlated with his adaptation in the soul's capacities — (a) cognitive faculties, (b) feeling of pleasure or displeasure, and (c) faculty of desire — of Tetens' trichotomy of feeling, understanding, will. (Teo, 2005). In terms of mathematics, Apostol in his book on calculus defined "The Law of Trichotomy" as: Every real number is negative, 0, or positive. The law is sometimes stated as "For arbitrary real numbers a and b, exactly one of the relations a < b, a = b, and a > b holds" (Apostol, 1967)" (Osler, 2012).

Understanding the Neutral Part of Trichotomy

Extensive exploration of "the neutral" and the concept of "neutrality" was explored extensively by mathematician and mathematical researcher Florentin Smarandache in the 1980s. Smarandache referred to his ideas regarding neutrality via the "Neutrosophic Philosophy" (a term he published in 1980 that is the philosophy of the neutral) and "Neutrosophics" which is "the science and study of neutrality" also founded by Smarandache during the same time period. Smarandache also founded and made practically applicable "Neutrosophic Logic" during his most productive research years. In terms of Triology and Trioengineering the "Neutrosophic Philosophy" is described by the author in terms of the triological neutral as an in-depth part of an intensive and comprehensive research investigation immersed in "Triological Logic" from a "Triostatistical Research Perspective" as follows:

"From the "Triological Perspective" the neutral part of neutral third part of the triological perspective of a trichotomy from the "Mathematical Law of Trichotomy". As such, the neutral is best defined as a "diverse" set or group of qualitative responses that contextually respond to the initial research question. Therefore, (from a mathematical, statistical, and inquiry perspective) the neutral part of the trichotomy as addressed by mathematical researcher Florentin Smarandache in his 2016 book entitled, "Neutrosophic Overset, Neutrosophic Underset, and Neutrosophic Offset" (a title which is in fact trichotomous in nature). Smarandache is the originator of "Neutrosophic Theory" that began as a scientific discipline via his research publications on the subject that began in earnest in 1980".

The Development of Trioengineering as In-Depth Analytics via the Triological Expansion of the Base Mathematical Law of Trichotomy into the Extensive Triology Provided by the 3-4-5-6 GURT

The Trioengineering 3-4-5-6 GURT has a fivefold trichotomy that together create the in-depth **"Trioengineering Quanta-Trichotomy"** are listed respectively as follows:

1.) The Inter-Trichotomy;
2.) The Intra-Trichotomy;
3.) The Meta-Trichotomy;
4.) The Para-Trichotomy; and lastly
5.) The Quadra-Trichotomy.

The fivefold trichotomies inherent to the 3-4-5-6 GURT are as the **"Trioengineering Quanta-Trichotomy"** (that is a Quinta-Trichotomy or "Trichotomy in five aspects", thus the **"Trioengineering Quanta-Trichotomy"** = the **"Trioengineering Quinta-Trichotomy"**) are respectively defined as follows:

"Inter-Trichotomy"

Definition: "Inter-Trichotomy" is a novel term that is a hyphenated portmanteau of the prefix "Inter" (meaning "between", "among" or "in the midst of") and trichotomy (meaning "a measurement of three as defined by the "Mathematical Law of Trichotomy""). Altogether the hyphenated term describes the 3-4-5-6 GURT used in Trioengineering as trifold respective sides a, b, and c that make up the complete GURT model and are used in the Triostatistics "[TEM]" (or "Triangular Equation Modeling") to represent the final Trioengineering trichotomous trifold three aspects or particular parameters to a given solution.

"Intra-Trichotomy"

Definition: "Intra-Trichotomy" is a novel term that is a hyphenated portmanteau of the prefix "Intra" (meaning "within") and trichotomy (meaning "a measurement of three as defined by the "Mathematical Law of Trichotomy""). Altogether the hyphenated term describes the 3-4-5-6 GURT used in Trioengineering as the trifold respective angles: (1. *alpha* ("α" = "angle alpha" = "∠α" = 36.87°); (2.) *rho* ("ρ" = "angle rho" ="∟ρ" = 90°); and lastly (3.) *beta* ("β" = "angle beta" = "∠β" = 53.13°) that make up the complete GURT model and are used in the Triostatistics "[TEM]" (or "Triangular Equation Modeling") to represent the final Trioengineering trichotomous trifold three specific aspects to a given solution represented "within the GURT" (often written outside of the respective GURT identified angle for convenience and clarity) as each of the respective GURT angles (that relate to each other in the [TEM] in a counterclockwise cyclical rotational diagram.

"Meta-Trichotomy"

Definition: "Meta-Trichotomy" is a novel term that is a hyphenated portmanteau of the prefix "Meta" (meaning "beyond") and trichotomy (meaning "a measurement of three as defined by the "Mathematical Law of Trichotomy""). Altogether the hyphenated term describes the 3-4-5-6 GURT used in Trioengineering as the comprehensive meaning of the 3-4-5-6 GURT beyond its simplicity as an upright right geometric model into its more complex problem-solving solution identified by the model's utility as the Triostatistics [TEM].

"Para-Trichotomy"

Definition: "Para-Trichotomy" is a novel term that is a hyphenated portmanteau of the prefix "Para" (meaning "beside" and "alongside of") and trichotomy (meaning "a measurement of three as defined by the "Mathematical Law of Trichotomy""). Altogether the hyphenated term describes the 3-4-5-6 GURT used in Trioengineering as the trifold respective sides *a*, *b*, and *c* along with the respective GURT angles alpha (("α" = "angle alpha" = "∠α" = 36.87°), rho ("ρ" = "angle rho" ="∟ρ" = 90°) , and beta ("β" = "angle beta" = "∠β" = 53.13°) that make up the complete GURT model and are used in the Triostatistics [TEM]. The 3-4-5-6 GURT takes into account that its respective sides and angles work alongside each other as a collective trichotomous whole to create problem-solving solutions that are illustrated and exhibited via the Triostatistics [TEM].

"Quadra-Trichotomy"

Definition: "Quadra-Trichotomy" is a novel term that is a hyphenated portmanteau of the prefix "Quadra" (literally meaning the number "four") and trichotomy (meaning "a measurement of three as defined by the "Mathematical Law of Trichotomy""). Altogether the hyphenated term describes the 3-4-5-6 GURT used in Trioengineering as the trifold respective sides *a*, *b*, and *c* that make up the complete GURT model and have the following unit measurements: Side *a* = 3; Side *b* = 4; Side *c* = 5; and lastly the Overall unit measurement of the Area of the entire GURT = 6.

The Seven Specific Types of Human Brainwaves and Brainwave Patterns

Although the brain is always emitting brainwaves at each different band it is the dominant brainwaves at a particular time that dictate the conscious state of an individual. By using audio technology, they can be studied. Audio technology can determine the bandwidth frequency of human brainwaves to specifically determine their unique properties and exact frequencies measured in Hertz ("*hz*"). It is important to note that brainwaves control all of the functions of the human body and some very exact frequencies have been found to cause specific observable effects in the brain such as the release of neurotransmitters and hormones identified as Serotonin (or "Human Growth Hormone").

Learning Domain	Frequency in Hertz	Visible Light Color	Physiological Association	Representative Icon	Comprehensive Definitions
Alpha [A] Brainwave Pattern	Cognitive	Operates Between 8hz – 12hz	Blue (Primary Color)	The Head	Alpha brainwaves are associated with learning, mental states, and generating creative ideas and problem-solving solutions. Any time there is an insight or an inspiration, the brain just produces more Alpha waves than usual.
Beta [B] Brainwave Pattern	Psychomotor	Operates Between 12hz – 38hz	Yellow (Primary Color)	The Hand	Beta brainwaves are the ones associated with normal waking consciousness. Low amplitude beta with multiple and varying frequencies often associated with active, busy or anxious thinking and active concentration (and can help with focused concentration, alertness and increasing learning)
Theta [Θ] Brainwave Pattern	Affective	Operates Between 3hz – 8hz	Red (Primary Color)	The Heart	Theta brainwaves have been identified as the gateway to feelings and emotions (it has been noted that the highest amounts of Human Growth Hormone (or HGH) are released at high Theta frequency and the Earths ionosphere resonates to the same rhythm). This frequency is also where much of the brains normally unused areas become most active.
Alpha + Beta = [A] + [B] = Gamma [Γ] Brainwave Pattern	Cognitive – Psychomotor = The Metacognitive Learning Domain	Operates Between 40hz – 100hz	Blue + Yellow = Green (Secondary Color)	The Head + The Hand	Gamma brainwaves are a very important frequency when it comes to higher awareness (and spiritual experiences).
Alpha + Theta = [A] + [Θ] = Lambda [Λ] Brainwave Pattern	Cognitive – Affective = The Ultraffective Learning Domain	Operates Between 100hz – 200hz	Blue + Red = Purple (Secondary Color)	The Head + The Heart	Lambda brainwaves very high frequency brainwaves. Associated with wholeness and integration (also associated with spiritual experiences). Interestingly these extremely high frequency brainwaves seem to ride on a very low frequency Epsilon wave.
Beta + Theta = [B] + [Θ] = Epsilon [E] Brainwave Pattern	Psychomotor – Affective = Neuromotive Learning Domain	Operates at Less Than > 0.5hz	Yellow + Red = Orange (Secondary Color)	The Hand + The Heart	Epsilon brainwaves have been measured to be below 0.5hz. They are strongly related to the highest frequency Lambda brainwaves in that if you zoom in far enough you would see them embedded within the slow Epsilon frequency is a very fast Lambda frequency wave. The same states of consciousness are associated with both Lambda and Epsilon waves. Wholeness and integration seem to be the main themes of these brainwaves.
Delta [Δ] = Sleeping Brainwave Pattern (as Dormant Brainwave Functionality)	All Learning Domains at Rest	Operates Between 0.5hz – 3hz	Visible Light not Perceived by the Human Eye because it is at Rest	The Entire Human Body at Rest (Sleep)	Delta is the adult sleeping brain wave. Delta activity is characterized by frequencies under Hz and is absent in awake healthy adults, but is physiological and normal in awake children under the age of 13. Delta waves are also naturally present in stage three and four of sleep (deep sleep) but not in stages 1, 2, and rapid eye movement (REM) of sleep.

TRIOENGINEERING ™ © *The Problem-Solving Triological Science: The In-Depth Trichotomous Science of the Dynamic 3-4-5-6 Golden Upright Right Triangle for Innovative Problem-Solving.* Osler Studios Incorporated ©, © Copyright 2022 All Rights Reserved.

The Trioengineering Six Learning Sciences Directly Derived from the Six Triological Trioengineering Learning Domains

The Triological Six Learning Domains (Three Primary and Three Secondary)	The Triological Six Learning Sciences (from the Six Learning Domains)
Cognitive	**Perceptology** (from Visualus) = The In-Depth Study of the Cognitive Learning Domain. Perceptology is the Triological Trioengineering Tripositional: **[Science of "*Comprehension*"]**
Affective	**Perspectology** (from Visualus) = The In-Depth Study of the Affective Learning Domain. Perspectology is the Triological Trioengineering Tripositional: **[Science of "*Disposition*"]**
Psychomotor	**Psychomotology** = The In-Depth Study of the Psychomotor Learning Domain. Psychomotology is the Triological Trioengineering Tripositional: **[Science of "*Action*"]**
Metacognitive	**Metacogtology** = The In-Depth Study of the Metacognitive Learning Domain. Metacognitive is the Triological Trioengineering Tripositional: **[Science of "*Higher-Order Comprehension*"]**
Ultraffective	**Ultraffectology** = The In-Depth Study of the Ultraffective Learning Domain. Ultraffectology is the Triological Trioengineering Tripositional: **[Science of "*Higher-Order Disposition*"]**
Neuromotive	**Neuromotology** = The In-Depth Study of the Neuromotive Learning Domain. Neuromotive is the Triological Trioengineering Tripositional: **[Science of "*Higher-Order Action*"]**

Perceptology (from Visualus) = **[Science of "*Comprehension*"]** = Comprehension as defined here can be trichotomously defined as the field of: "Understanding"; "Overstanding"; or "Outerstanding" equivalent to the following identities: "Understanding" ≡ "Learning About"; "Overstanding" ≡ "Mastery Over"; or "Outerstanding" ≡ "Disconnected From".

Perspectology (from Visualus) = **[Science of "*Disposition*"]** = Disposition as defined here can be trichotomously defined as the arena of emotions conveyed or expressed as: "Attitudes"; "Opinions"; or "Feelings" that can be trichotomously measured as: "Positive" ≡ "[+]"; "Negative" ≡ "[–]"; or "Neutral" ≡ "[ø]".

Psychomotology = **[Science of "*Action*"]** = Action as defined here can be trichotomously defined as actions that are expressed and can be trichotomously measured as: "Positive" ≡ "[+]"; "Negative" ≡ "[–]"; or "Neutral" ≡ "[ø]".

Metacogtology = **[Science of "*Higher-Order Comprehension*"]** = Higher-Order Comprehension (which is the summative combination of the "Cognitive" + "Psychomotor" primary learning domains to create this higher-order secondary learning domain and is therefore the summative combination of the sciences of "Perceptology" + "Psychomotology").

Ultraffectology = **[Science of "*Higher-Order Disposition*"]** = Higher-Order Disposition (which is the summative combination of the "Cognitive" + "Affective" primary learning domains to create this higher-order secondary learning domain and is therefore the summative combination of the sciences of "Perceptology" + "Perspectology").

Neuromotology = **[Science of "*Higher-Order Action*"]** = Higher-Order Action (which is the summative combination of the "Affective" +

"Psychomotor" primary learning domains to create this higher-order secondary learning domain and is therefore the summative combination of the sciences of "Perspectology" + "Psychomotology").

Chapter 10 follows and describes the Optimization as it is applied via Trioengineering.

TRIOENGINEERING ™ © *The Problem-Solving Triological Science: The In-Depth Trichotomous Science of the Dynamic 3-4-5-6 Golden Upright Right Triangle for Innovative Problem-Solving.* Osler Studios Incorporated ©, © Copyright 2022 All Rights Reserved.

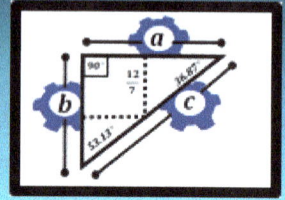

To everything there is a season, and a time to every purpose under the heaven.

Ecclesiastes 3: 1

The Origins of Optimal Instruction

"Optimal Instruction" is the terminology for the empowering art of instruction pioneered and developed by the author to describe a broad–based and strategically designed teaching methodology that empowers students by allowing them to: (1) identify of their gifts and talents; (2) determine their purpose; (3) draw upon their experiences; and (4) increase their knowledge–base in the instructional setting. The operational definition of "Optimal Instruction" is as follows:

Broadly defined "Optimal Instruction" is "The strategy implemented during instruction to achieve optimal learning". "Optimal Instruction" or "Optimizing Instruction" is the art of strategically unlocking a learner's unlimited potential through a dynamic and engaging instructional strategy. Optimal Instruction or "OI" is grounded in engaging content delivery methods, the process of instructional design, and Metacognetic problem–solving methodologies. Specifically, *Optimal Instruction is the art of analyzing and authoring, designing and developing, engaging and implementing an information delivery system that enlightens and empowers the learner.*

An "**Ergonomic Information Architect ©**" is one who uses the process of Optimal Instruction. The goal of the Ergonomic Information

Architect (or **"EIA"**) is to create an empowering learning environment that is relevant and allows the learner to blend their individual skills, gifts, talents, and experiences with the course content to promote higher order thinking and content mastery.

The Optimal Instruction Implementation Methodology: The Application of Visualus Constructed Concepts and Techniques to Teaching

The Optimal Instruction Implementation Methodology that drives the instructional content delivery and facilitation of the retention and transfer of content knowledge in the learning environment was formulated from many years of teaching and research on Instructional Design, Distance Learning, and Effective Content Delivery methodologies. Optimal Instruction is the dynamic instructional content delivery strategy which incorporates the Metacognetic Mechanics discipline and its practice of Technology Engineering to augment and enhance the learning environment. The primary idea is to create an educational setting in which content delivery is transparent, enjoyable, and empowering, thereby, producing seamless content mastery. The elements of Optimal Instruction are as follows:

The Elements of Optimal Instruction

1. Activate the Audience's Level of Interest with Exciting Measures:

Capture the audiences' attention by using or developing instructional techniques that capture interest, are engaging, activate curiosity, elicit experience, and have relevance to learners. Engage Metacognetic Mechanics' principles through the use of Technology Engineered tools at any point to create a dynamic and exciting "positive cognitive economy of self–motivated learners".

2. Gather Data from the Audience on Negative A Priori Experiences:

Elicit from each learner their "A Priori" (or "before") Curricular Experiences on the course content to determine their level of knowledge regarding the content and to determine if they have preconceived fears, notions or anxieties about the subject matter. Acquire permission and share the experiences with the entire class through active learner–centered discourse to build a common bond, create help–centered teams, and promote an overall sense of community.

3. Provide Ongoing Positive Reinforcement:

Reward, Acknowledge, and Encourage at all times during the learning process. Always be positive and willing to share encouraging comments even when the situation appears to be negative to the learner or audience. This stage is the most important and most crucial. *A continually positive learning environment creates positive learners* and can change even the most hardened learner into an empowered conduit of knowledge. *Remember: Reward, Acknowledge, and Encourage at all times.*

4. Model Positive Desired Behaviors as an Instructor:

Model the behavior that you want to elicit from learners. Share your outcomes and demonstrate how to accomplish tasks with question-and-answer discussions. Deliver to the Audience the: What, When, Where, Why, How, and How Much that are required for all course-based assessments and assignments.

5. Record Individual Critical Reflections:

Require learners to record their Critical Reflections after every experience. This focuses student responses in the three domains of learning: Cognitive, Affective, and Psychomotor labeled as Knowledge, Skills, and Dispositions. This will provide the Instructor with greater insight as to how learners are progressing via their hopes, feelings, and abilities as they matriculate through the content.

6.   Establish Relevance: By Eliciting and Sharing of Audience A Priori and A Posteriori Experiences:

Elicit from each learner their A Priori (before) and A Posteriori (after) experiences via discussions about goals, relevant use of the content knowledge in their own lives to determine their level of expectations and feelings towards accomplishing tasks. Share the experiences with the entire class through active discourse to increase the sense of community.

7.   Progressive Mastery Learning:

Do not allow forward progression into new areas of content knowledge until learners have mastered the present content knowledge at 100%. Allow for repetition of content until it is mastered. Accept no less than perfect mastery and continually encourage at all times. Even if learners have to repeat the content 100 times, the repetition is excellent for them and successfully encodes the data into their minds. Soon learners will expect to achieve at higher levels and will push not only themselves but the entire group into higher levels of achievement.

The Major Components of Optimal Instruction

To initiate a successful teaching strategy often one must have a battery of logical and effective instructional systems, instructional tools, and instructional strategies. The following is a list of the five components vital to the process of Optimal Instruction as a teaching strategy.

Immediately following the list is a comprehensive narrative describing each component in detail:

1. **Ergonomics;**
2. **Information Architecture ©;**
3. **Ergonomic Information Architecture © (or "EIA ©");**
4. **The Application of Ergonomic Information Architecture; and**
5. **Ergonomic Information Architecture and the Learner.**

Ergonomics

Ergonomics can be broadly defined as "Human Centered Engineering" or more generally as "Usability Engineering". Specifically, the term "Ergonomics" comes from two Greek words: "Ergon" meaning "work" and "Nomoi" meaning "Natural Laws". An Ergonomist (a scholar, researcher or scientist who uses Ergonomics) literally studies human factors or environmental capabilities in direct relationship to work and work requirements or demands.

Information Architecture

"Information Architecture" is the practice of structuring and organizing information for a particular purpose. It is a unique combination of two terms developed to describe the process of organizing data digitally for a variety of users. The term has particular relevance for those who develop information for the Internet (web designers, web authors, graphic designers, and online developers). The term **"Information Architect"** is generally used to describe those professionals who design and organize models, procedures, and systems to help others assess, deliver, find, and manage information.

"Ergonomic Information Architecture ©"

Ergonomic Information Architecture (abbreviated as "EIA") is the cohesion of the two aforementioned practices into a new process. This new process is purely Metacognetic and involves the development of instructional methods, models, procedures, strategies, and systems that take into account human–centered factors that will enable others to more readily and easily: evaluate, deliver, develop, design, find, learn, manage, and organize information. Ergonomic Information Architecture is the combination of the two Optimal Instructional elements: **Ergonomics + Information Architecture**. By combining these two elements we take the unique strengths of both to create a means of developing Instructional tools that can be specifically developed to meet the needs of learners.

The EIA process is merely a descriptor for an "Instructional Scientist" who utilizes the discipline of Metacognetic Mechanics to develop Instructional Tools through the practice of Technological Engineering to increase the retention and transfer of content knowledge in the learning environment. The "Instructional Scientist" who uses this process is referred to as an **"Ergonomic Information Architect"**.

The Educational Application of Ergonomic Information Architecture

In education, teaching, and the learning environment, "ergonomic information architecture" or "[EIA]" can be broadly defined as "the application of specific instructional theories and techniques in a systematic way to enhance learner retention, transfer, and synthesis of content knowledge." It is a comprehensive and systematic way of completing all of the methods necessary to accomplish a specified task". As it applies to the delivery of knowledge, teaching evolves into "ergonomic instruction" through the practice of ergonomic information architecture. The result is strategic teaching techniques that place emphasis on creating instructional strategies that are specifically designed to aid learners in maximizing their performance while enabling them to relate the knowledge directly to their own experiences.

Ergonomic Information Architecture and the Learner

Many instructional techniques can be used to develop an ergonomic teaching strategy. project–based learning and learner–based tools are just two of the many examples. The goal is to produce learners that have not only mastered the course material, but also have learned to make the subject matter a relevant and active part of their personal goals and objectives. essentially, the learner is aided in the memorization of information and the implementation of the information in a personal and meaningful way. the overall idea is to empower the individual. In this manner, the instructor in the learning environment actively promotes learning in an interesting and vital way. this creates learners that are motivated, willing to initiate, and remain positive about their learning. thus, lifelong–learning is active and is actively promoted in the learning environment.

Developing an Effective Ergonomic Teaching Strategy through the Use of Ergonomic Information Architecture

The purpose of Ergonomic Information Architecture is to maximize learner performance and encourage self-motivated learning. Some key points that an effective Ergonomic Information Architect may keep in mind when developing a curriculum that uses the principles of EIA are:

1. Promote interest in the course subject matter by making the knowledge relevant to the learner.
2. Develop doable, concise, comprehensive, consistent, and specific learner objectives.

3. Build upon the learner's prior knowledge and experiences by making content knowledge relevant in practical ways.
4. Encourage teamwork and collaboration whenever possible.
5. Provide consistent and steady positive reinforcement by providing ongoing encouragement.
6. Develop a "High Locus of Learner Control" by providing assignments or projects that allow the learner to be in control of their learning through procedures that are self–paced and draw upon their unique individual experiences.
7. Continually refine and improve the learning environment and teaching methodologies through consistent formative and summative evaluation.

Some alternative methods that can be used to develop strong and effective ergonomic teaching strategies are:

1. Efforts that encourage continual collaboration.
2. Techniques that allow for student empowerment.
3. Strategies that take advantage of shared resources.
4. Ongoing curriculum development with peers and other professionals.
5. Built-in techniques and strategies for problem analysis and problem–solving solutions.
6. Efforts that take advantage of opportunities for instructional improvement through experiences that encourage professional training and development.
7. Strategies and methods that develop creative problem–solving via the implementation of brainstorming tools and strategies.
8. Integrating Technology Engineering tools into the curriculum in dynamic, useful, and interesting ways.
9. Implementing methods and strategies that take advantage of individual learner innovation and creativity.

10. Team building projects, assignments, and tasks that allow for equality in team member input and efforts.

Technology Engineering and Optimal Instruction in Face to Face and Distance Education University Undergraduate and Graduate Statistics Courses

Previous work by the author involved the development of effective Interactive Metametric Tools and their infusion into course content. Inquiry in this arena led to the primary research question: "Is it possible to combine dynamic instructional strategies with subject matter and interactive tools in such a way that there is an increase in subject matter retention and transfer by students?" Results of this research exceeded expectations. Outcomes included comments that were lively, engaging, and unexpectedly empowering.

A new methodology for instruction had been created that established a powerful "Interactive Cognitive Economy" (an "Interactive Cognitive Economy" takes advantage of collaboration, meaningful relevance to the learner regarding the subject matter, challenging and engaging learning activities, and meaningful performance objectives and goals). In a Technologically Engineered Interactive Cognitive Economy students become active learners who are consciously engaged in developing their own education. Ultimately, this new teaching methodology combined the principles of instructional design with interactivity, discovery, relevance, collaboration, creativity, and innovation. This new instructional methodology was dubbed: "Technology Engineering".

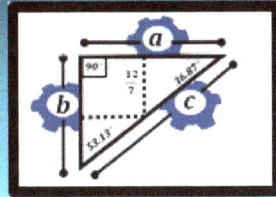

And the rain descended, and the floods came, and the winds blew, and beat upon that house; and it fell not: for it was founded upon a rock.

Matthew 7: 25

The "Trioengineering Thought" Holistic Universal Application of Trioengineering Concepts, Models, and Expressions

Information from the Trioengineering E-Book entitled, "Trioengineering Thought" published by the author.

11

The Trioengineering Methodology: Trichotomy Defined (Part One)
Seeing the Mathematical Law of Trichotomy in Nature

The Subatomic Structure of the Atom is Naturally Trichotomous. Note the following model:

One begins with observing Trichotomy in nature. The Subatomic structure of the "Atom" which is the building block for of material existence, it has:

1.) A Proton (which is a Positive Particle = "+");
2.) An Electron (which is a Negative Particle = "−"); and lastly
3.) A Neutron (which is a Neutral Particle = "ø").

The Proton and Neutron together makeup the atom's nucleus (in its center) and the Nucleus is orbited by the Electron. Observe: →

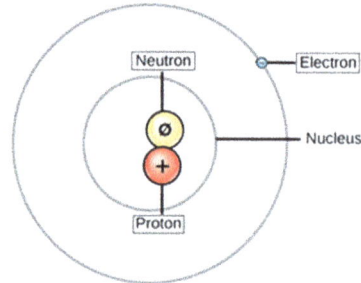

Note: The Subatomic structure of the "Atom" which is the building block for of material existence, it has a: Proton (Positive); Electron (Negative); and the Neutron (Neutral).

16

The Trioengineering Methodology (continued): The Trichotomous Neuroscience of Decision-Making

The Trichotomous Trioengineering Methodology is instinctively used in the human brain (which is in fact multiplicatively trichotomous in its inner functions; inner workings; and overall infrastructure) when one has to make a decision about some particularly identified subject matter.

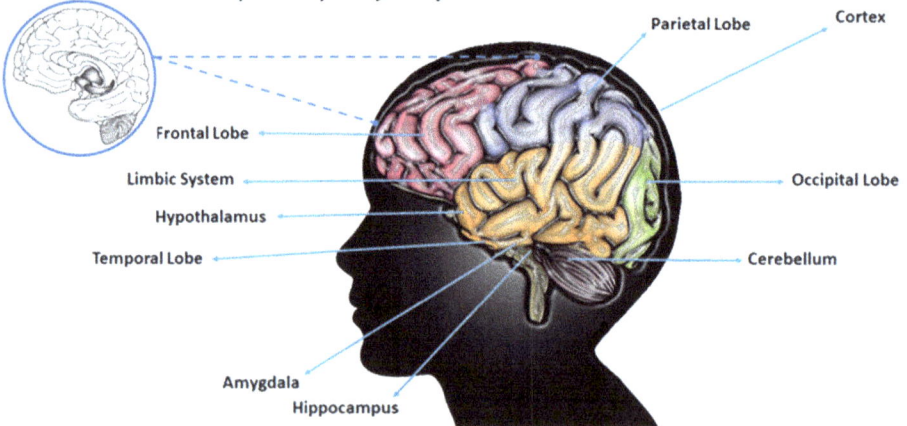

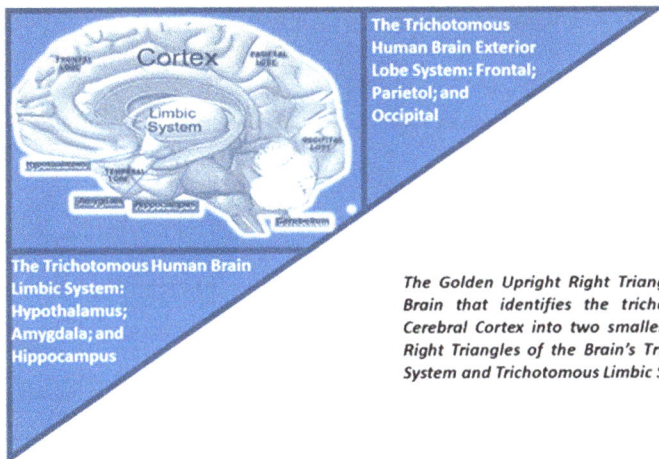

The Trioengineering Trichotomous Modeling: The Trichotomous Neuroengineering of the Human Brain

The Trichotomous Human Brain Exterior Lobe System: Frontal; Parietol; and Occipital

The Trichotomous Human Brain Limbic System: Hypothalamus; Amygdala; and Hippocampus

The Golden Upright Right Triangle Model of the Human Brain that identifies the trichotomous regions of the Cerebral Cortex into two smaller more definitive Upright Right Triangles of the Brain's Trichotomous Exterior Lobe System and Trichotomous Limbic System).

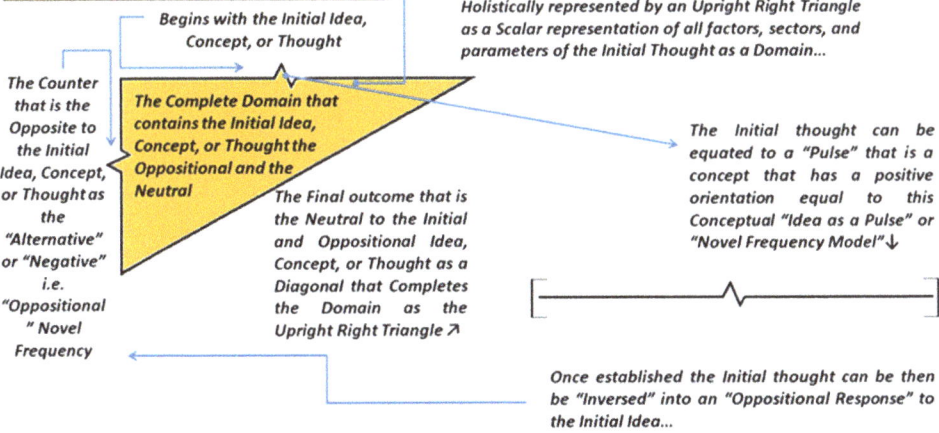

The Trioengineering Methodology: An Initial Thought that Transitions into Trichotomous Conceptualizations

Begins with the Initial Idea, Concept, or Thought

Conceiving of the Initial Thought as an Idea that is in fact a "Complete Domain" from which can portioned into: A Positive, A Negative, and A Neutral... Holistically represented by an Upright Right Triangle as a Scalar representation of all factors, sectors, and parameters of the Initial Thought as a Domain...

The Counter that is the Opposite to the Initial Idea, Concept, or Thought as the "Alternative" or "Negative" i.e. "Oppositional" Novel Frequency

The Complete Domain that contains the Initial Idea, Concept, or Thought the Oppositional and the Neutral

The Final outcome that is the Neutral to the Initial and Oppositional Idea, Concept, or Thought as a Diagonal that Completes the Domain as the Upright Right Triangle ↗

The Initial thought can be equated to a "Pulse" that is a concept that has a positive orientation equal to this Conceptual "Idea as a Pulse" or "Novel Frequency Model" ↓

Once established the Initial thought can be then be "Inversed" into an "Oppositional Response" to the Initial Idea...

TRIOENGINEERING ™ © *The Problem-Solving Triological*

Science: The In-Depth Trichotomous Science of the Dynamic 3-4-5-6 Golden Upright Right Triangle for Innovative Problem-Solving. Osler Studios Incorporated ©, © Copyright 2022 All Rights Reserved.

The Trioengineering Methodology (continued): Creating The Total Trioengineering Model

31

The Initial Idea: [+]   Next→   **The Oppositional Idea: [–]**   and Lastly→   **The Neutral Idea: [Ø]**

Written on this line is the initial Idea; Concept; or Thought.

Written on this line is the oppositional Idea; Concept; or Thought.

Written on this line is the neutral Idea; Concept; or Thought that is neither the 1st or 2nd outcome.

Illustrating The Total Trioengineering Model Growth:

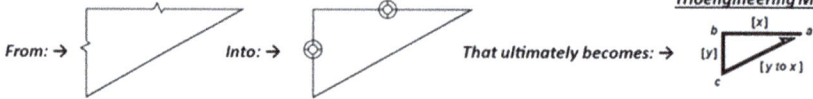

The Total Trioengineering Model:

From: →   *Into: →*   *That ultimately becomes: →*

$$[y] \quad \begin{array}{c} b \quad [x] \quad a \\ \\ c \quad [y \text{ to } x] \end{array}$$

Within The Trioengineering Methodology:

[x] = The Initial Idea: Next→   **[y] = The Oppositional Idea:**   and Lastly→   **[y to x] = The Neutral Idea**

Written on this line is the initial Idea; Concept; or Thought.

[Upright Right Triangle Side a.]

Written on this line is the oppositional Idea; Concept; or Thought.

[Upright Right Triangle Side b.]

Written on this line is the neutral Idea; Concept; or Thought that is neither the 1st or 2nd outcome.

[Upright Right Triangle Side c.]

The Trioengineering Methodology (continued): Model Measurement and Expansion in Inverse

32

From a Conceptual Model   *Into a Targeted Area: →*

That has the two designated primary target areas: ↗

Creating the
← *Overall Upright Triangle Model →*

With One Single Neutral Area based on Targets 1 & 2: ↑

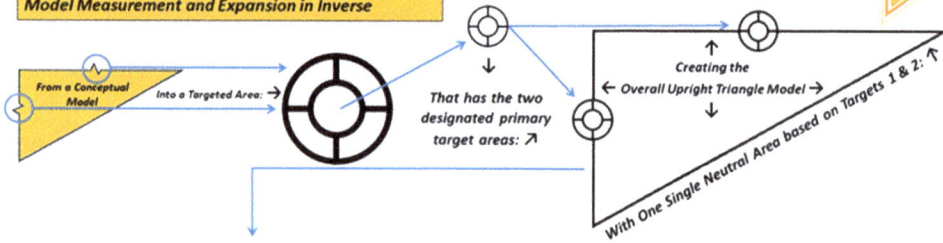

That becomes the Golden Upright Right Triangle:
↓

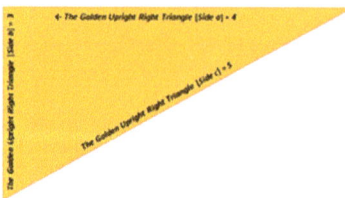

← *The Golden Upright Right Triangle [Side a] = 4*

The Golden Upright Right Triangle [Side c] = 5

← *The Golden Upright Right Triangle [Side b] = 3*

That has the following measurements: →

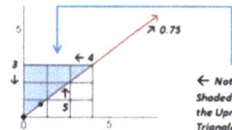

↗ 0.75

← *Note: The Shaded Area that is the Upright Right Triangle with its respective Side Measurements on the Graph.*

TRIOENGINEERING ™ © *The Problem-Solving Triological Science: The In-Depth Trichotomous Science of the Dynamic 3-4-5-6 Golden Upright Right Triangle for Innovative Problem-Solving.* Osler Studios Incorporated ©, © Copyright 2022 All Rights Reserved.

The Trioengineering Methodology (continued): Mathematical Measurement of the Golden Upright Right Triangle

Measurements in terms of the Pythagorean Theorem:

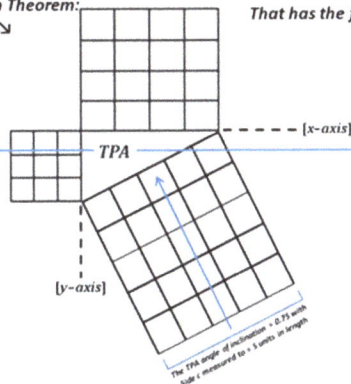

TPA

[x-axis]

[y-axis]

The TPA angle of inclination = 0.75 with Side c measured to = 5 units in length

That has the following meaning in the following measurements:

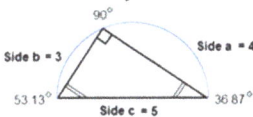

$$\text{Incline: } m = \left[\frac{\Delta y}{\Delta x}\right] \text{ where } \left[\begin{array}{l} = 3/4 = 0.75 = \text{to the inclination or} \\ \text{the diagonal of the triangle, because } \Delta x \\ = \text{Side } a = 4 \text{ and } \Delta y = \text{Side } b = 3. \end{array}\right]$$

also

Measurements in terms of Thales Theorem:

$90°$

Side b = 3   Side a = 4

$53.13°$ Side c = 5 $36.87°$

TPA = is the acronym for "Trichotomous Progression Analysis" – Which is a Triostatistical measure of growth via a line from a base line to an upper region, in terms of the Trioengineering Model, TPA is generally measured via the diagonal Upright Right Triangle Line with a 0.75 inclination (indicating strong growth on a 0.00 to 1.00 scale).

Trioinformatics Trioengineering Neuroengineering Defined through Neuroscience and Neuromathematics Tabular Terminology Symbols that Apply to the Eight Area Definitions

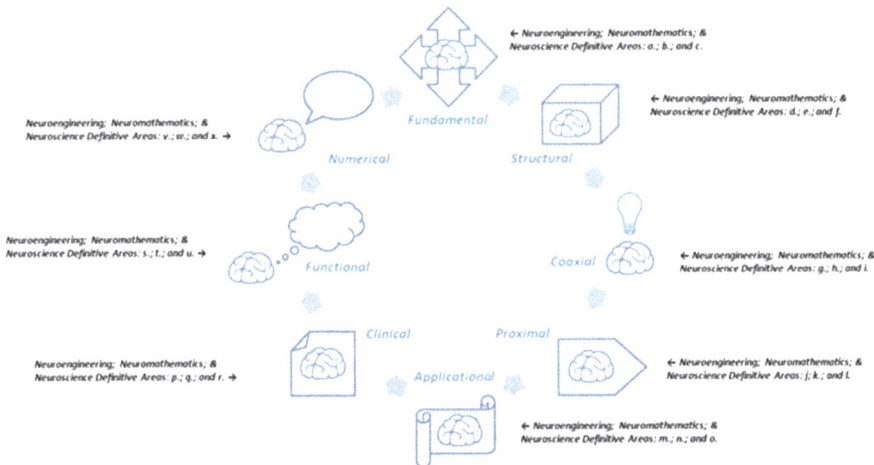

← *Neuroengineering; Neuromathematics; & Neuroscience Definitive Areas: a.; b.; and c.*

← *Neuroengineering; Neuromathematics; & Neuroscience Definitive Areas: d.; e.; and f.*

Neuroengineering; Neuromathematics; & Neuroscience Definitive Areas: v.; w.; and x. →

← *Neuroengineering; Neuromathematics; & Neuroscience Definitive Areas: g.; h.; and i.*

Neuroengineering; Neuromathematics; & Neuroscience Definitive Areas: s.; t.; and u. →

Neuroengineering; Neuromathematics; & Neuroscience Definitive Areas: p.; q.; and r. →

← *Neuroengineering; Neuromathematics; & Neuroscience Definitive Areas: j.; k.; and l.*

← *Neuroengineering; Neuromathematics; & Neuroscience Definitive Areas: m.; n.; and o.*

Fundamental   *Numerical*   *Structural*   *Functional*   *Coaxial*   *Clinical*   *Proximal*   *Applicational*

Trioengineering Mathematics: All Positive Integers are Right Triangle Numbers

54

- Initially Defined: By Mathematician Carl Gauss in his:
- Original "Eureka! Triangle Number Theorem" = "$n(n + 1) \div 2$";
- Has been proven true in nature in the Mathematical Law of Trichotomy as related to the subatomic particle—atom basic components;
- Also Proven True in Pascal's Triangle;
- Extended by the author in his Upright Right Triangle Number Theorem; and
- Is the foundation of Mathematical Trioengineering and support by the author's discovery and establishment of the field of Triostatistics.

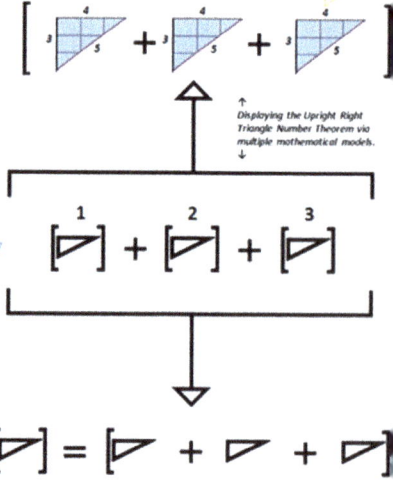

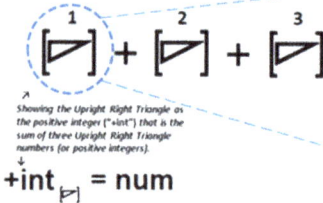

Displaying the Upright Right Triangle Number Theorem via multiple mathematical models.

Showing the Upright Right Triangle as the positive integer ("+int") that is the sum of three Upright Right Triangle numbers (or positive integers).

$+int \; \triangleright = num$

$num = \triangleright = \triangleright + \triangleright + \triangleright$

Base Trioengineering Mathematical Models for Trioengineering Mathematical Research

58

The measurements for the final Trioengineering outcome are all based of off the Golden Upright Right Triangle which is the base Triostatistical model for all Trioengineering-related concepts, methods, techniques, and solutions.

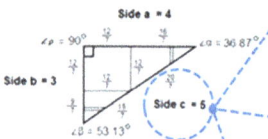

$\angle \rho = 90°$

Side a = 4

Side b = 3

Side c = 5

$\angle a = 36.87°$

$\angle B = 53.13°$

Note: How the measurements are derived and are all balanced coming from the same Golden Upright Right Triangle and have merit via Trioengineering.

Override: Side c "Construction" Operations

$$c = b\left[\Delta[x]\right] \equiv b\left[\sqrt{1 + \left[\frac{a}{b}\right]^2}\right]$$

Provides the length or 'distance measure' of the Side c which is equal to the 'Contrapositive' of the Trichotomous Progression Line within the confines of the Upright Right Triangle which is one half of the Tri-Squared Analysis 3 by 3 Table Format and the front of the Pauwlus Isometric Cuboid. This particular equation and identification is referred to as the 'Inclination Identification Equation'.

The In-Depth descriptions of the Trichotomous Progression Line, Pauwlus Isometric Cuboid Front Side: the Triangle Equation Modeling Upright Right Triangle and the Tri-Squared Analysis 3 by 3 Table.

• The positive direction of the Trichotomous Line

(x_2, y_2)

a — The Override

b — The Underside

(x_1, y_1) — The Outside

$\frac{\Delta y}{\Delta x} = \left[\frac{b}{a}\right]$

$$\frac{\Delta y}{\Delta x} = \left[\frac{b}{a}\right]$$

As the difference in the Sides a and b of the Upright Right Triangle to calculate the Trichotomous Progression Line

The unchanging height of the positive line as inclination

The elongated Trichotomous Progression Line illustrated with the 3 by 3 Tri-Squared Analysis 3 by 3 Table creating the precise measurements for the Upright Right Triangle within the 3 by 3 Table creating the front of the Pauwlus Isometric Cuboid

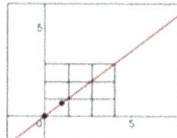

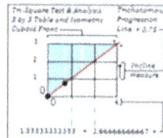

Tri-Square Test & Analysis 3 by 3 Table and Isometric Cuboid Paper

Trichotomous Progression Line = 0.75

An example of the Underside equivalent measures for the inclination of the TPA Zone or Upright Right Triangle:

Underside measurement examples:

(x_2, y_2)

(x_1, y_1)

Incline: $m = \frac{\Delta y}{\Delta x}$

Note: For the 3-4-5 TPA Zone (Upright Right Triangle) measures:

$m = 0.75$

$b = (0.2)$ thus $a = b$

$y = ma + b = 0.75(a) + b$

$\theta = 53.13 + \tan(\theta)$

Inclination: $m = \frac{\Delta y}{\Delta x} = \tan(\theta)$

TRIOENGINEERING ™ © *The Problem-Solving Triological Science:* The In-Depth Trichotomous Science of the Dynamic 3-4-5-6 Golden Upright Right Triangle for Innovative Problem-Solving. Osler Studios Incorporated ©, © Copyright 2022 All Rights Reserved.

Base Trioengineering Mathematical Models for Trioengineering Mathematical Research

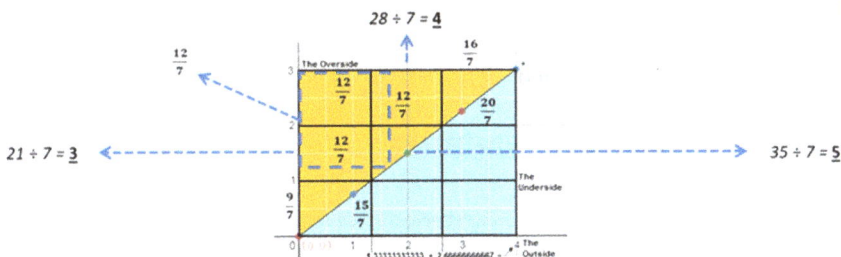

$$28 \div 7 = \underline{4}$$

$\dfrac{12}{7}$

$21 \div 7 = \underline{3}$

$\dfrac{16}{7}$

$35 \div 7 = \underline{5}$

From The Golden Upright Right Triangle into The Tri-Squared Test Standard 3 by 3 Table Format:

- **The Golden Upright Right Triangle:** As the initial measurement area;
- **The Inverted Golden Upright Right Triangle:** Opposite of the initial Golden Upright Right Triangle;
- **The Combined Golden Upright Right Triangle + The Inverted Golden Upright Right Triangle:** Into a unified new whole (as illustrated above); and
- **Creating the Final Tri-Squared Test Standard 3 by 3 Table Format.**

The 4A Metric Algorithm Standards ©

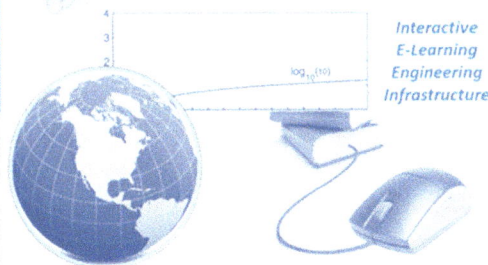

Interactive E-Learning Engineering Infrastructure

TRIOENGINEERING ™ © *The Problem-Solving Triological*

Science: The In-Depth Trichotomous Science of the Dynamic 3-4-5-6 Golden Upright Right Triangle for Innovative Problem-Solving. Osler Studios Incorporated ©, © Copyright 2022 All Rights Reserved.

THE 4A METRIC STANDARDS

Description of the Standards

The 4A Metric Algorithm Standards ©, are a set of four standards aimed specifically at the use and integration of technology in the online teaching and learning environment. They are also applicable to other modalities of instruction such as in more traditional and dual (also known as hybrid) educational settings. The Standards are grounded in the ideology espoused and posited by the Academy of Press Educators that focuses on learner self-growth as well as the measurement of learning from novice to mastery initially presented by Dr. James E. Osler in the 2010 book entitled, "Infometrics ©" (where the 4A Metric was first introduced). The 4A Metric Standards set precedence as they are the first set of rigorously designed guiding practices for digital learning that are designed to simultaneously mathematically measure learner creativity and growth (as systemically assessed by both the student and the teacher). Holistically, the four comprehensive 4A Metric Standards are designed to aid in the diverse practice of planning, designing, and developing the architecture of courses for the E-Learning environment.

Chapter Twelve follows and provides examples of how to conduct a rapid Trioengineering Calculation starting with models that illustrate the entire solution creation and tangible outcome production process.

The stone which the builders rejected, the same is become the head of the corner: this is the Lord's doing, and it is marvelous in our eyes.

Matthew 21: 42

The Detailed Information Provided by the E-Book Entitled: "The 4A Metric Algorithm Standards"© Published by the Author

Information and images on the pages that follow were extracted from the Trioengineering E-Book entitled, "The 4A Metric Algorithm Standards ©".

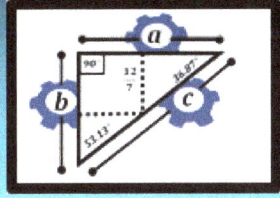

TRIOENGINEERING ™ © *The Problem-Solving Triological*

Science: The In-Depth Trichotomous Science of the Dynamic 3-4-5-6 Golden Upright Right Triangle for Innovative Problem-Solving. Osler Studios Incorporated ©, © Copyright 2022 All Rights Reserved.

THE 4A METRIC® STANDARDS

Standard One: Structure

In Standard One "Structure" is defined as: "The systemic functionality and use of the 4A Metric Algorithm © to deliver subject matter content in a sequential scaffolding format that both empowers and engages learners. The use of the innovative components and elements of the Algorithm allow it move the learner from a subject matter "novice level" to a final "master level" by the end of the course through a series of in-depth systemically assigned progressive experiences that provide the learner with a "high locus of control" and the ability to consistently and constantly "self-assess their learning". The instructor uses the sequential procedures of the metric to create a course in one of two methods: a.) Use of the 4A Metric Algorithm Infrastructure as a "Structurally Static" "Instructional Educational Ecosystem" (or "Edusystem"); or b.) Use the 4A Metric Algorithm Infrastructure as a "Highly Flexible" "Instructional Educational Ecosystem" (or "Edusystem")."

THE 4A METRIC® STANDARDS

The image below is a sample of the published article that supports Standard One.

CASE STUDY

THE 4A METRIC ALGORITHM: A UNIQUE E-LEARNING ENGINEERING SOLUTION DESIGNED VIA NEUROSCIENCE TO COUNTER CHEATING AND REDUCE ITS RECIDIVISM BY MEASURING STUDENT GROWTH THOROUGH SYSTEMIC SEQUENTIAL ONLINE LEARNING

Standard One: Structure
Supportive Reference

Reference: Osler, J. E. (2016). The 4A Metric Algorithm: A Unique E-Learning Engineering Solution Designed via Neuroscience to Counter Cheating and Reduce Its Recidivism by Measuring Student Growth through Systemic Sequential Online Learning. i-manager's September–November Journal of Educational Technology, 12 (2), pp. 44–61.

TRIOENGINEERING ™ © *The Problem-Solving Triological Science: The In-Depth Trichotomous Science of the Dynamic 3-4-5-6 Golden Upright Right Triangle for Innovative Problem-Solving.* Osler Studios Incorporated ©, © Copyright 2022 All Rights Reserved.

THE 4A METRIC® STANDARDS

Standard Two: Presentation

Standard Two "Presentation" is defined as: "The actual appearance and exhibition of the course subject matter content using 4A Metric Algorithm © infrastructure (as either "static" or "flexible") in a sequential scaffolding format that is designed to empower and engage learners as they progress from the "novice level" to the "master level". The components and elements of the Algorithm allow it to present the subject matter content in different formats that take advantage of "Multiple Learning Styles" and the "Multiple Intelligences". Learning through the systemic and sequential 4A Metric course design allows for self-paced learning that is highly interactive. The instructor takes advantage of presentation system to maximize learner subject matter experiences by utilizing the following: a.) The Course "Learning Management System" (or "LMS") unique features and functions to sculpt into the elements, components, and levels of the 4A Metric ©; b.) Multimedia to deliver multiple forms of subject matter content to stimulate interest and meet the unique learning needs of all learners completing the course; and c.) A course-based dynamic "Virtual E-Portfolio" as an Experience Application Programming Interface (or "Experience API") to provide a high learner locus of control by creating an authentic course archive that is also a detailed "Digital Learner Self-Assessment System" that records learner growth and supports learner success through sequentially detailed micro and macro course credentialing."

THE 4A METRIC® STANDARDS

Standard Two: Presentation
Supportive Reference

The image below is a sample of the published article that supports Standard One.

Reference: Osler, J. E. & Wright, M. A. (2016). Neuro–Holistic Learning ©: An Integrated Kinesthetic Approach to Cognitive Learning © using Collaborative Interactive Thought Exchange © in a Blended Environment to Enhance the Learning of Young African American Males. i-manager's January–March Journal of Educational Technology, 12 (4). pp. 1–9.

Standard Three: Modality

Standard Three "Modality" is defined as: "The delivery of the course and the subsequent course subject matter content to t[] learner using the 4A Metric Algorithm © infrastructure using the following methodology: a.) As "Synchronous" (meaning real tir[] instructor and learner interaction via recorded video through teleconferencing software); b.) As "Asynchronous" (meaning provisi[] of 4A Metric © course subject matter and associated resources at all times through the course LMS); and c.) Modality defined [] one of three functional ways to provide the 4A Metric © course instruction to the learner as—(1.) "Online Instruction" (meaning th[] the course is provided 100% of the time through the course LMS), (2.) "Traditional Instruction" (meaning that the course [] provided through "face to face" instructor to learner "traditional classroom interaction"), and (3.) "Dual Instruction" (meaning th[] the course is taught as "Blended or Hybrid Instruction" which is "face to face traditional classroom interaction" with simultaneo[] support from online instruction through the course LMS)."

The 4A Metric °
Standards Clearinghouse °

THE **4A** METRIC °

Standard Three: Modality
Supportive Reference

Reference: Osler, J. E. & Wright, M. A. (2015). Dynamic Neuroscientific Systemology: Using Tri–Squared Meta–Analysis and Innovative Instructional Design to Develop a Novel Distance Education Model for the Systemic Creation of Engaging Online Learning Environments. i-manager's July–September Journal of Educational Technology, 12 (2), pp. 42–55.

The image below is a sample of the published article that supports Standard One

The 4A Metric °
Standards Clearinghouse °

THE **4A** METRIC °

THE 4A METRIC ® STANDARDS

Standard Four: Measurement

Standard Four "<u>Measurement</u>" is defined as: "The use of the 4A Metric Algorithm © procedures to assess and evaluate learner growth from novice level to master level. In terms of measurement, the 4A Metric © uses internal repetitive "Mastery Testing" that is naturally ADA compliant and designed to ensure mastery before sequentially proceeding to the next of the Metric's 4 Levels of subject matter expertise. The metric also has a unique and innovative series of sequential systemic point systems that is mathematically calculable and tied to micro and macro credentialing. These innovative and unique mathematical procedures are all used in the course infrastructure and are listed as follows: a.) The 4A Metric Algorithm Regular Quadrilateral Function of E-Learning © (used to determine the efficacy of the 4A Metric instructional methodology); b.) The 4A Metric Algorithm Point Scale © (used to determine the specific point total of each individual item and requirement used throughout the 4A Metric designed course); c.) The 4A Metric Virtual E-Portfolio Rubric Grading Scale © (used to determine the numerical grading scale for items submitted in the self-growth evaluation course archive Virtual E-Portfolio); d.) Authentic workforce development rewards as micro and macro credentials (designed to enable the learner to be immediately marketplace ready and viable); e.) The 4A Metric 4 Levels © which are unique methods used to measure learner self-growth from novice to mastery based upon the "Taxonomy of Process Education"; and f.) Triostatistics (through statistical measures such as the "Tri-Squared Test" used to determine and further measure the instructional and learning efficacy of the 4A Metric © through in-depth research inquiry)."

THE 4A METRIC ® STANDARDS

Standard Four: Measurement
Supportive Reference

The image below is a sample of the published article that supports Standard One

Reference: Osler, J. E. (2017). An Innovative Psychometric Model for Algorithmic Repetitive Mastery Testing. i-manager's February–April Journal on Educational Psychology, 10 (4), pp. 19–28.

TRIOENGINEERING ™ © The Problem-Solving Triological

Science: The In-Depth Trichotomous Science of the Dynamic 3-4-5-6 Golden Upright Right Triangle for Innovative Problem-Solving. Osler Studios Incorporated ©, © Copyright 2022 All Rights Reserved.

The Comprehensive Triangular Trichotomous
Model of the "Metadomains of Learning" Venn Diagram©

The Osler Triangular Trichotomous Venn Diagram that Defines the Metadomains of Learning ©

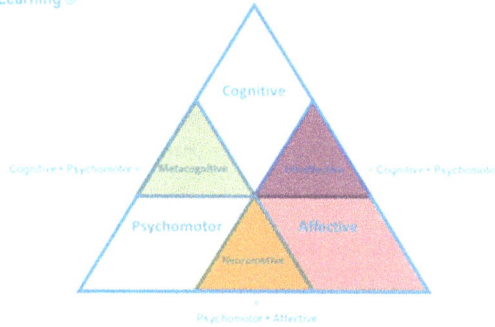

The Comprehensive Triangular Trichotomous
Model of the "Metadomains of Learning" Venn Diagram©

The Osler Triangular Trichotomous Venn Diagram that Defines the Neuroscience of the Three Metadomains of Learning ©

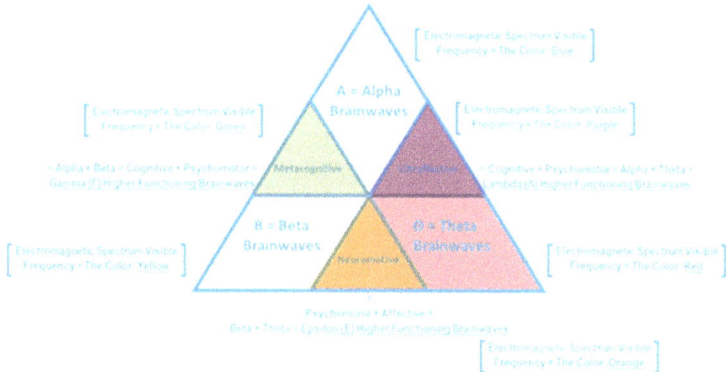

TRIOENGINEERING ™ © *The Problem-Solving Triological*

Science: The In-Depth Trichotomous Science of the Dynamic 3-4-5-6 Golden Upright Right Triangle for Innovative Problem-Solving. Osler Studios Incorporated ©, © Copyright 2022 All Rights Reserved.

The Comprehensive Credentialing Models of the 3 Learning Domains©

The Osler Trichotomous Shape Models that Defines the Credentialing for the Three Learning Domains ©

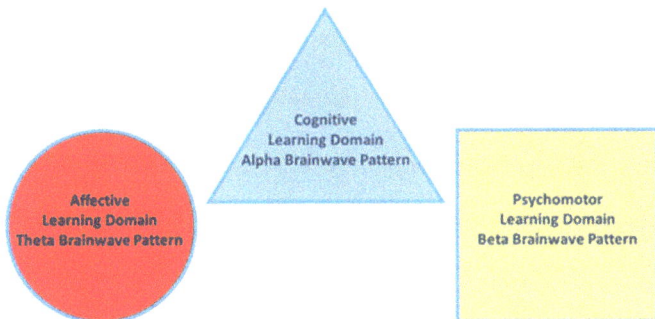

Cognitive Learning Domain Alpha Brainwave Pattern

Affective Learning Domain Theta Brainwave Pattern

Psychomotor Learning Domain Beta Brainwave Pattern

The 4A Metric ®
Standards Clearinghouse ®

THE **4A** METRIC

The Comprehensive Credentialing Model of the 3 Learning Metadomains©

The Osler Triangular Trichotomous Credentialing Shapes that Define the Neuroscience of the Three Learning Metadomains ©

The Ultraffective ©
Learning Domain
(Lambda Brainwave Pattern)

The Metacognitive ©
Learning Domain
(Gamma Brainwave Pattern)

The Neuromotive ©
Learning Domain
(Lambda Brainwave Pattern)

The 4A Metric ®
Standards Clearinghouse ®

THE **4A** METRIC

TRIOENGINEERING ™ © *The Problem-Solving Triological*

Science: The In-Depth Trichotomous Science of the Dynamic 3-4-5-6 Golden Upright Right Triangle for Innovative Problem-Solving. Osler Studios Incorporated ©, © Copyright 2022 All Rights Reserved.

COGNITIVE Learning Domain Competencies as Credential Categories:

CRITIQUE   INQUIRY   SYSTEMS

The 4A Metric *
Standards Clearinghouse *

THE 4A METRIC STANDARDS

Copyright * James E. Osler II Ed
2018 All Rights Reserv

COGNITIVE Learning Domain Competencies as Credential Categories:

EXPERIENCE   OPERATIONS

The 4A Metric *
Standards Clearinghouse *

THE 4A METRIC STANDARDS

Copyright * James E. Osler II Ed
2018 All Rights Reserv

TRIOENGINEERING ™ © The Problem-Solving Triological

Science: The In-Depth Trichotomous Science of the Dynamic 3-4-5-6 Golden Upright Right Triangle for Innovative Problem-Solving. Osler Studios Incorporated ©, © Copyright 2022 All Rights Reserved.

AFFECTIVE *Learning Domain Competencies as Credential Categories:*

DISPOSITION

COMPASSION

EMPATHY

The 4A Metric *
Standards Clearinghouse *

THE **4A** METRIC *STANDARDS*

Copyright * James E. Osler II Ed.D.
2018 All Rights Reserved.

AFFECTIVE *Learning Domain Competencies as Credential Categories:*

RESPONSIVE

CARING

The 4A Metric *
Standards Clearinghouse *

THE **4A** METRIC *STANDARDS*

Copyright * James E. Osler II Ed.D.
2018 All Rights Reserved.

TRIOENGINEERING ™ © *The Problem-Solving Triological*

Science: The In-Depth Trichotomous Science of the Dynamic 3-4-5-6 Golden Upright Right Triangle for Innovative Problem-Solving. Osler Studios Incorporated ©, © Copyright 2022 All Rights Reserved.

PSYCHOMOTOR *Learning Domain Competencies a* Credential Categories:

COLLABORATE

COMMUNICATE

CONSTRUCT

The 4A Metric*
Standards Clearinghouse*

THE 4A METRIC* STANDARDS

Copyright* James E. Osler II Ed|
2018 All Rights Reserv|

PSYCHOMOTOR *Learning Domain Competencies a* Credential Categories:

UTILIZE

PROTOTYPE

The 4A Metric*
Standards Clearinghouse*

THE 4A METRIC* STANDARDS

Copyright* James E. Osler II Ed|
2018 All Rights Reserv|

TRIOENGINEERING ™ © *The Problem-Solving Triological*

Science: The In-Depth Trichotomous Science of the Dynamic 3-4-5-6 Golden Upright Right Triangle for Innovative Problem-Solving. Osler Studios Incorporated ©, © Copyright 2022 All Rights Reserved.

METACOGNITIVE Learning Domain Competencies as Credential Categories:

CAREER

ADAPTIVE

PLANNING

METACOGNITIVE Learning Domain Competencies as Credential Categories:

IDENTIFY

IDEATION

TRIOENGINEERING ™ © *The Problem-Solving Triological*

Science: The In-Depth Trichotomous Science of the Dynamic 3-4-5-6 Golden Upright Right Triangle for Innovative Problem-Solving. Osler Studios Incorporated ©, © Copyright 2022 All Rights Reserved.

ULTRAFFECTIVE *Learning Domain Competencies* a
Credential Categories:

MEASURE INVESTIGATE RESEARCH

The 4A Metric *
Standards Clearinghouse *

THE 4A METRIC *

Copyright * James E. Osler II E
2018 All Rights Reser

ULTRAFFECTIVE *Learning Domain Competencies* a
Credential Categories:

VIRTUOSITY ANALYZE

The 4A Metric *
Standards Clearinghouse *

THE 4A METRIC *

Copyright * James E. Osler II Ed.D
2018 All Rights Reserved

TRIOENGINEERING ™ © *The Problem-Solving Triological Science: The In-Depth Trichotomous Science of the Dynamic 3-4-5-6 Golden Upright Right Triangle for Innovative Problem-Solving.* Osler Studios Incorporated ©, © Copyright 2022 All Rights Reserved.

NEUROMOTIVE Learning Domain Credential Categories:

PROBLEM-SOLVING

DISCOVERY

EXPERIMENT

The 4A Metric®
Standards Clearinghouse®

THE **4A** METRIC® STANDARDS

Copyright® James E. Osler II Ed.D.
2018 All Rights Reserved.

NEUROMOTIVE Learning Domain Credential Categories:

PERSERVERANCE

CREATIVITY

The 4A Metric®
Standards Clearinghouse®

THE **4A** METRIC® STANDARDS

Copyright® James E. Osler II Ed.D.
2018 All Rights Reserved.

TRIOENGINEERING ™ © *The Problem-Solving Triological*

Science: The In-Depth Trichotomous Science of the Dynamic 3-4-5-6 Golden Upright Right Triangle for Innovative Problem-Solving. Osler Studios Incorporated ©, © Copyright 2022 All Rights Reserved.

Final Summary

The author firmly believes that the formal study and application of Trioengineering as an in-depth Triological Science will greatly aid in the production of globally beneficial problem–solving solutions. Through Trioengineering, many different types of scientists as: educators, investigators, instructors, inventors, and researchers now have a definitive universal problem–solving methodology that will allow them to be more productive, and have access to practical solutions that can be made applicable to virtually any tangible and digital situation. In terms of education, Trioengineers now have an in-depth operational methodology that will allow them to provide impacting, empowering, and inspirational instructional delivery methods as a thorough and comprehensive solution to measurable online learning. This is the ultimate outcome desired by the Trioengineer and is the ideal outcome sought by those who ultimately use Trioengineering as a Triological scientific discipline.

This concludes **TRIOENGINEERING**™© *The In-Depth Trichotomous Science of the Dynamic 3-4-5-6 Golden Upright Right Triangle for Innovative Problem-Solving*.

I know thy works, and charity, and service, and faith, and thy patience, and thy works; and the last to be more than the first.

Revelation 2: 19

Behold, how good and how pleasant it is for brethren to dwell together in unity!
Psalm 133: 1

A History of Published Work by the Author that Formed the Foundation for TRIOENGINEERING ™ ©

Osler, J. E. (1996). The Effects Of An Ergonomically Designed Computer–Based Tutorial On Elementary Students' Recall. Raleigh, NC: College of Education and Psychology – North Carolina State University.

Osler, J. E. (1996). A Mathematical Equation Expressing the Rectilinear Propulsion of the 100 and 110 Meter Hurdle Races © (Research Report). Durham, NC.

Osler, J. E. (2002). University Management Develop Program Comprehensive Report.

Osler, J. E. (2004). Dimensions of Teaching and Behavioral Impediments of Teaching Efficacy. Ideas About Teaching Efficacy: Sharing Perspectives. National Social Sciences Press. Gabe Keri.

Osler, J. E. (2004). The Crisis: Classroom Culture, Identifying and Analyzing Seven Factors That Disable An Effective Collegiate Teaching Methodology. A Long Way to Go: Conversations About Race By African American Faculty And Students. Peter Lang.

Osler, J. E. (2005) Technology Engineering: A Paradigm Shift In The Dynamics of Instruction; A New Philosophy of Education For Teaching In The Information Age. 2005 The South Atlantic Philosophy of Education Society Refereed 50th Conference.

Osler, J. E. (2005). Creating An Interactive Cognitive Economy: The Use of an Asynchronous Learning Network Course Management System to Develop an Interactive Community of Learners. Research Paper for 2005 The South Atlantic Philosophy of Education Society Refereed 50th Conference.

Osler, J. E. (2005). Technology Engineering: Developing, Implementing, and Infusing Interactive Metametric Learning Modules into an Asynchronous Learning Network to Develop an Interactive Community of Learners. Research Paper for The 2005 South Atlantic Philosophy of Education Society Refereed 50th Conference.

Osler, J. E. (2008). Σimply Σtatistics ©. A Comprehensive Guide to Statistical Formulae, Methodology, and Techniques. Durham, NC: Publishing Division, Osler Studios Incorporated™ ©.

Osler, J. E. (2009). Σimply Σtatistics ©. The Handheld Edition. Durham, NC: Publishing Division, Osler Studios Incorporated™ ©.

Osler, J. E. (2009). The Osler Micro–Lending Strategy ©. Durham, NC: Research Division, Osler Studios Incorporated™ ©.

Osler, J. E. (2010). PERCEPTOLOGY™ © The Science of Comprehension that is Universal Instructional Design through Visualus, Metacognetic Mechanics, Technology Engineering, and Optimal Instruction. Durham, NC: Publishing Division, Osler Studios Incorporated™ ©.

TRIOENGINEERING ™ ©

TRIOENGINEERING ™ © The Problem-Solving Triological Science: The In-Depth Trichotomous Science of the Dynamic 3-4-5-6 Golden Upright Right Triangle for Innovative Problem-Solving is an informative guidebook developed to explain in detail the innovative design and mathematical processes involved in the formation of relevant real-world solutions. TRIOENGINEERING derives its solutions from an intense in–depth comprehensive mathematical engine that is constructed with three–dimensional mathematics that define the application of the Innovative Problem–Solving Model of Inventive Instructional Design. A "Trioengineer" (a Problem–Solver who uses TRIOENGINEERING) is able to analyze, design, develop, implement, conduct formative and summative evaluation of tangible values designed to be beneficial solutions to real-world problems. TRIOENGINEERING has been written in a manner that deconstructs and defines the inventive process of Instructional Design as mathematical formulae, models, procedures, and terms while aiding the reader to understand how to use this innovative and unique technique to produce direct answers to any problem.

Published by: Osler Studios Incorporated ™. Durham, North Carolina

TRIOENGINEERING ™ © *The Problem-Solving Triological Science: The In-Depth Trichotomous Science of the Dynamic 3-4-5-6 Golden Upright Right Triangle for Innovative Problem-Solving.* Osler Studios Incorporated ©, © Copyright 2022 All Rights Reserved.

TRIOENGINEERING ™ © *The Problem-Solving Triological Science: The In-Depth Trichotomous Science of the Dynamic 3-4-5-6 Golden Upright Right Triangle for Innovative Problem-Solving.* Osler Studios Incorporated ©, © Copyright 2022 All Rights Reserved.

A Final Dedication to Almighty GOD

At the completion of this book, I wish to once again acknowledge and thank Almighty GOD. May this book be a blessing to those who read it and may it aid them in the completion of each of their respective endeavors and make the world a better place. In the blessed and holy name of my most precious Lord and Savior – Jesus Christ, AMEN.

Yours in Love, Truth, and Service

James E. Osler II, Ed.D.

"I thank God, whom I serve from my forefathers with pure conscience, that without ceasing I have remembrance of thee in my prayers night and day."

2nd Timothy 1: 3

The following statement is a Trioengineering ™ © affirmation:

"May Almighty GOD be Glorified by the work that you have given us to fulfill the Purpose and Destiny that you Created in each of us by your Divine and Holy Will."

"To Almighty GOD be the Glory!"

In Jesus name, Amen.

TRIOENGINEERING
THE PROBLEM-SOLVING TRIOLOGICAL SCIENCE ©

www.ingramcontent.com/pod-product-compliance
Lightning Source LLC
Chambersburg PA
CBHW060317100426
42812CB00003B/805